Lecture Notes in Mathematics

1666

Editors:
A. Dold, Heidelberg
F. Takens, Groningen

T0225925

Springer
Berlin
Heidelberg
New York
Barcelona
Budapest
Hong Kong
London
Milan
Paris
Santa Clara
Singapore
Tokyo

Marius van der Put
Michael F. Singer

Galois Theory
of Difference Equations

 Springer

Authors

Marius van der Put
Department of Mathematics
University of Groningen
P.O. Box
NL-9700 AV Groningen, The Netherlands
e-mail: M.van.der.Put@math.rug.nl

Michael F. Singer
Department of Mathematics
North Carolina State University
Box 8205, Raleigh, N.C. 27695-8205, USA
e-mail: singer@math.ncsu.edu

Cataloging-in-Publication Data applied for

Die Deutsche Bibliothek - CIP-Einheitsaufnahme

Put, Marius van der:
Galois theory of difference equations / Marius van der Put ; Michael
F. Singer. - Berlin ; Heidelberg ; New York ; Barcelona ; Budapest ;
Hong Kong ; London ; Milan ; Paris ; Santa Clara ; Singapore ;
Tokyo : Springer, 1997
　(Lecture notes in mathematics ; 1666)
　ISBN 3-540-63243-3

Mathematics Subject Classification (1991): 12H10, 39A10

ISSN 0075-8434
ISBN 3-540-63243-3 Springer-Verlag Berlin Heidelberg New York

Typesetting: Camera-ready T$_E$X output by the authors
SPIN: 10553306　　46/3142-543210 - Printed on acid-free paper

Contents

Algebraic Theory

We shall develop here a Galois theory for difference equations. In the Picard-Vessiot Galois theory for *differential* equations, the basic objects of study are the *Picard-Vessiot extension* of a field and its associated Galois group. Recall that a *Picard-Vessiot extension* of a difference field k is an extension K generated by a fundamental set of solutions of a linear differential equation and having the same field of constants as k. The automorphism group of K over k is an algebraic group and properties of this group reflect properties of the differential equation. When one tries to mimic this approach for difference equations one is confronted with the following example (recall that a difference field is a field k together with an automorphism ϕ of k):

Example 0.1 ([25]) *Let k be a difference field whose characteristic is not 2. If the field of constants $C_k = \{c \in k \mid \phi(c) = c\}$ is algebraically closed, then the equation $\phi x + x = 0$ has no nonzero solution in k. To see this note that if $y \in k$ satisfies $\phi y + y = 0$ then $\phi(y^2) = y^2$ so y^2 is a constant. Since C_k is algebraically closed, we have y is a constant, contradicting the fact that $\phi y = -y$.*

Therefore, if one restricts oneself to fields, the properties of having algebraically closed constants and having full sets of solutions of difference equations can be incompatible. A field theoretic Galois theory for difference equations was developed by Franke [25] who investigated its ramifications. The main deficiency of this theory is that one could not associate a Picard-Vessiot-type extension to every difference equation. We take a different approach. Fahim [23], Levelt [36], and van der Put [49] showed that a Galois theory of differential equations could be based on *rings*, in particular, simple differential rings. In the differential case, the rings are integral domains and so have quotient fields which are the Picard-Vessiot extensions. We shall follow this approach and develop a theory based on simple difference rings. These rings will be shown to be reduced but can have zero divisors. Nonetheless they are the natural analogue of Picard-Vessiot extensions.

In Chapter 1, we will develop the basic properties of the Picard-Vessiot theory of difference equations - their existence and unicity, and the existence of a Galois group that is a linear algebraic group. In Chapter 2, we discuss algorithms for determining the Galois group of difference equations of order 1 and difference equations in diagonal form. We also outline the algorithm of Hendriks [26] for determining the Galois group of second order difference equations. Chapter 3 is devoted to giving an algebraic (and constructive) proof of the fact that every connected linear algebraic group is the Galois group of a Picard-Vessiot difference ring over $C(z)$, where C is an arbitrary algebraically closed field of characteristic zero and the difference operator is defined by $\phi(z) = z + 1$. We return to this question in Chapter 8 where we show (using analytic tools) that a necessary and sufficient condition for a linear algebraic group to be the Galois group of a difference equation over $\mathbf{C}(\{z^{-1}\})$, ($\mathbf{C}$ being the complex numbers) is that G/G^0 is cyclic, where G^0 is the connected component of the identity in G. Chapter 4 applies the Galois theory of difference equations to the study of algebraic properties of linear recursive and differentially finite sequences, that is, sequences satisfying difference equations over C and $C(z)$ respectively. Chapter 5 considers difference equations over $\overline{\mathbf{F}}_p(x)$, where $\overline{\mathbf{F}}_p$ is the algebraic closure of the field with p elements. We show that there is a simple classification of difference modules M over this field. Furthermore, we show that the difference Galois group of M is the Zariski closure (over $\overline{\mathbf{F}}_p(x^p - x)$) of the cyclic group generated by the "p-curvature of M". We also compare the characteristic zero and characteristic p theories and show that the natural analogue of the Grothendieck conjecture is false for difference equations. In Chapter 6, we give the classification of difference modules over $\mathcal{P}$, the field of Puiseux series in $t = x^{-1}$, where $\phi(t^{1/m}) = t^{1/m}(1 + t)^{-1/m}$. The results here are similar to the formal local classification of differential modules and form the starting point for the study of analytic properties of difference modules.

We wish to thank Anne Duval for many discussions during the preliminary stages of writing and in particular for her help with the material in Chapters 6 and 7.

Chapter 1

Picard-Vessiot rings

We begin this section with several definitions.

Definition 1.1   *1. A difference ring is a commutative ring R, with 1, together with an automorphism $\phi : R \to R$. If, in addition, R is a field, we say that R is a difference field.*

   2. The constants of a difference ring R, denoted by C_R are the elements $c \in R$ satisfying $\phi(c) = c$.

   3. A difference ideal of a difference ring is an ideal I such that $\phi(a) \in I$ for all $a \in I$. A simple difference ring is a difference ring R whose only difference ideals are (0) and R.

Example 1.2 Let $\mathbf{C}$ be the field of complex numbers. Each of the fields

- $\mathbf{C}(z)$, the field of rational functions in z,

- $\mathbf{C}(\{z^{-1}\})$, the fraction field of convergent power series in z^{-1},

- $\mathbf{C}((z^{-1}))$, the fraction field of formal power series in z^{-1},

are all difference fields with ϕ given by $\phi(z) = z + 1$. For the last two fields this means that ϕ is given by $\phi(t) = \frac{t}{1+t}$ where $t = z^{-1}$. Note that this automorphism extends to

- $\mathcal{P}$, the algebraic closure of $\mathbf{C}((z^{-1}))$, which is also called the field of the formal Puiseux series,

by putting $\phi(t^{\frac{1}{m}}) = t^{\frac{1}{m}}(1+t)^{-\frac{1}{m}}$. ∎

Example 1.3 Consider the set of sequences $\mathbf{a} = (a_0, a_1, \ldots)$ of elements of an algebraically closed field C. We define an equivalence relation on this set by saying that two sequences $\mathbf{a}, \mathbf{b}$ are equivalent if there exists an N such that $a_n = b_n$ for all $n > N$. Using coordinatewise addition and multiplication, one sees that the set of such equivalence classes forms a ring $\mathcal{S}$. The map $\phi_0((a_0, a_1, a_2, \ldots)) = (a_1, a_2, \ldots)$ is well defined on equivalence classes (one needs to work with equivalence classes to have the property that this map is injective). The ring $\mathcal{S}$ with the automorphism ϕ_0 is therefore a difference ring. To simplify notation we shall identify an element with its equivalence class. The field C may be identified with the subring of constant sequences $(c, c, c, \ldots)$ of $\mathcal{S}$. If the characteristic of C is zero then any element of $C(z)$ is defined for sufficiently large integers (note that in characteristic p, this is not true for $(z^p - 1)^{-1}$). Therefore the map $f \mapsto (f(0), f(1), \ldots)$ defines a difference embedding of $C(z)$ into $\mathcal{S}$. Note that the map the map $f \mapsto (f(0), f(1), \ldots)$ also defines a difference embedding of $\mathbf{C}(\{z^{-1}\})$ into $\mathcal{S}$.

We note that $\mathcal{S}$ is not a simple difference ring. To see this let $\mathbf{a}$ be any sequence whose support (i.e., those integers i such that $a_i \neq 0$) is an infinite set of density zero in the integers (e.g., $\mathbf{a} = (a_i)$ where $a_i = 1$ if i is a power of 2 and 0 otherwise). The ideal generated by $\mathbf{a}, \phi_0(\mathbf{a}), \phi_0^2(\mathbf{a}), \ldots$ is a nontrivial difference ideal in $\mathcal{S}$. ∎

Let R be a difference ring. For $A \in Mat_n(R)$

$$\phi Y = AY$$

denotes a first order linear difference system. We shall restrict ourselves to equations where $A \in Gl_n(R)$ (to guarantee that we get n independent solutions). Here Y denotes a column vector $(y_1, \ldots, y_n)^T$ and $\phi Y = (\phi y_1, \ldots, \phi y_n)^T$. Given an nth order difference equation $L(y) = \phi^n y + \ldots + a_1 \phi Y + a_0 Y = 0$ we can consider the equivalent system

$$
\begin{pmatrix} \phi y \\ \phi^2 y \\ \vdots \\ \phi^n y \end{pmatrix} = \begin{pmatrix} 0 & 1 & 0 & \ldots & 0 \\ 0 & 0 & 1 & \ldots & 0 \\ \ldots & \ldots & \ldots & \ldots & \ldots \\ -a_0 & -a_1 & \ldots & -a_{n-2} & -a_{n-1} \end{pmatrix} \begin{pmatrix} y \\ \phi^1 y \\ \vdots \\ \phi^{n-1} y \end{pmatrix}
$$

For this system, the condition that the matrix lies in Gl_n is that $a_0 \neq 0$.

Definition 1.4 *Let R be a difference ring and $A \in Gl_n(R)$. A fundamental matrix with entries in R for $\phi Y = AY$ is a matrix $U \in Gl_n(R)$ such that $\phi U = AU$. If U and V are fundamental matrices for $\phi Y = AY$, then $V = UM$ for some $M \in Gl_n(C_R)$ since $U^{-1}V$ is left fixed by ϕ.*

Definition 1.5 *Let k be a difference field and $\phi Y = AY$ a first order system $A \in Gl_n(k)$. We call a k−algebra R a Picard-Vessiot ring for $\phi Y = AY$ if:*

1. An automorphism of R, also denoted by ϕ, which extends ϕ on k is given.

2. *R is a simple difference ring.*

3. *There exists a fundamental matrix for $\phi Y = AY$ with coefficients in R.*

4. *R is minimal in the sense that no proper subalgebra of R satisfies the conditions 1,2 and 3.*

We will show in the next section that if C_k is algebraically closed then, for any system $\phi Y = AY$, there is a Picard-Vessiot ring for this system and that it is unique up to $k-$difference isomorphism.

Example 1.6 Let C be an algebraically closed field of characteristic not equal to 2. R be the difference subring of $\mathcal{S}$ generated by C and $\mathbf{j} = (1, -1, 1, -1, \ldots)$. Note that $R = C[(1, -1, 1, -1, \ldots)]$. The 1×1 matrix whose only entry is $(1, -1, 1, -1, \ldots)$ is the fundamental matrix of the equation $\phi_0 y = -y$. This ring is isomorphic to $C[X]/(X^2 - 1)$ whose only non-trivial ideals are generated by the cosets of $X - 1$ and $X + 1$. Since the ideals generated in R by $\mathbf{j} + 1$ and $\mathbf{j} - 1$ are not difference ideals, R is a simple difference ring. Therefore R is a Picard-Vessiot extension of C. Note that R is reduced but not integral. ∎

In the following sections we will make use of the next elementary lemma.

Lemma 1.7 *a) The set of constants in a simple difference ring forms a field.*
b) If I is a maximal difference ideal of a difference ring R, then I is a radical ideal and for any $r \in R$, $\phi(r) \in I$ if and only if $r \in I$. Therefore R/I is a reduced difference ring.

Proof: a) If c is a constant, then $c \cdot R$ is a nonzero difference ideal so there is a $d \in R$ such that $c \cdot d = 1$. A computation shows that d is a constant.

b) To prove the first claim, one can easily show that the radical of a difference ideal is a difference ideal. To prove the second claim, note that $\{r \in R \mid \phi(r) \in I\}$ is a difference ideal that contains I but does not contain 1. ∎

The remainder of this section is organized as follows. In section 1.1 we show the existence and uniqueness of Picard-Vessiot rings assuming that the field of constants of k is algebraically closed. In section 1.2, we shall show that the group G of $k-$difference automorphisms of a Picard-Vessiot ring R that is a separable extension of k has the structure of an algebraic group over C_k and that R is the coordinate ring of variety which is a principal homogeneous space for G. In section 1.3 we consider the total quotient ring of a Picard-Vessiot ring and establish a Galois correspondence between certain difference subrings and closed subgroups of the Galois group. Finally in section 1.4, we consider the Tannakian category approach to defining the Galois group [20] and we will discuss the relation of our approach to this approach.

1.1 Existence and uniqueness of Picard-Vessiot rings

Let k be a difference field and let

$$\phi(Y) = AY \tag{1.1}$$

be a difference system with $A \in Gl(d)(k)$. To form a Picard-Vessiot ring for
(1.1) we proceed as follows. Let (X_{ij}) denote a matrix of indeterminates over k
and let det denote the determinant of this matrix. On the $k-$algebra $k[X_{ij}, \frac{1}{det}]$
one extends the automorphism ϕ be setting $(\phi X_{ij}) = A(X_{ij})$. If I is a maximal difference ideal of $k[X_{ij}, \frac{1}{det}]$ then Lemma 1.7 implies that $k[X_{ij}, \frac{1}{det}]/I$
is a simple difference ring. From the definition we see that $k[X_{ij}, \frac{1}{det}]/I$ is a
Picard-Vessiot ring for (1.1) and any Picard-Vessiot ring will be of this form. To
prove uniqueness of Picard-Vessiot rings, we need the following result. In this
result we restrict ourselves to difference fields with algebraically closed fields of
constants. This restriction excludes difference fields (k, ϕ) with ϕ of finite order.
In particular, $(\overline{\mathbf{F}}_p(z), \phi(z) = z + 1)$ is excluded.

Lemma 1.8 *Let R be a finitely generated k-algebra having an automorphism,
also called ϕ, extending ϕ on k. Let C be the constants of k and assume that
C is algebraically closed and that R is a simple difference ring. Then the set of
constants of R is C.*

Proof: Suppose that $b \notin C$ and $\phi(b) = b$. Consider the subring $C[b]$ of R.
Since R is simple every nonzero element f of this subring has the property
that $Rf = R$, i.e., there is an element $g \in R$ such that $fg = 1$. Since C
is algebraically closed it follows that $C[b]$ is a polynomial ring over C. Let $\overline{k}$
denote the algebraic closure of k. One sees that any nonzero element $f \in C[b]$
defines a regular, nowhere zero map of the affine variety $spec(\overline{k} \otimes_k R)$ to $\overline{k}$ whose
image is therefore a constructible subset of $\overline{k}$. Consider the map defined by
the element b. If $c \in C$ is in the image of this map then the map defined by
$b - c \in C[b]$ has a zero. Therefore the image of the map b has empty intersection
with C. It follows that the image of this map is finite and so there is a polynomial
$P = X^d + a_{d-1}X^{d-1} + ... + a_0 \in k[X]$ such that $k[b] = k[X]/(P)$. Since $\phi(b) = b$,
one finds that b also satisfies the polynomial $X^d + \phi(a_{d-1})X^{d-1} + ... + \phi(a_0)$.
The uniqueness of P implies that P lies in $C[X]$. This contradicts the fact that
$C[b]$ is a polynomial ring over C. ∎

Proposition 1.9 *Let k be a difference field with algebraically closed field of
constants and let R_1 and R_2 be Picard-Vessiot extensions of k for $\phi(Y) = AY$.
Then there exists a $k-$difference isomorphism between R_1 and R_2.*

Proof: We consider $R_1 \otimes_k R_2$ a difference ring where $\phi(r_1 \otimes r_2) = \phi(r_1) \otimes \phi(r_2)$.
Choose an ideal I in $R_1 \otimes_k R_2$ which is maximal in the collection of ϕ-invariant

ideals and put $R_3 = R_1 \otimes_k R_2/I$. The canonical maps $R_1 \to R_3$ and $R_2 \to R_3$ are injective since the kernels are ϕ-invariant ideals. The image of the first map is generated over k by a fundamental matrix in R_3 and similarly for the second map. Two fundamental matrices differ by a matrix with coefficients in C_{R_3}, which according to Lemma 1.8 is C_k. It follows that the two images are the same. Hence R_1 is isomorphic to R_2.   ∎

1.2   The Galois group

As an aid in understanding the structure of Picard-Vessiot rings, we will introduce a geometric point of view. As noted above, any Picard-Vessiot extension for (1.1) is of the form $k[X_{ij}, \frac{1}{det}]/I$ where I is a maximal ϕ–invariant ideal of $k[X_{ij}, \frac{1}{det}]$. Lemma 1.7 implies that such an ideal is a radical ideal and so is the ideal of a reduced algebraic subset of $Gl(d)_k = spec(k[X_{ij}, \frac{1}{det}])$. Let $\overline{k}$ denote the algebraic closure of k. The automorphism ϕ extends to an automorphism of $\overline{k}$ which will also be denoted by ϕ. The automorphism ϕ of $D := \overline{k}[X_{ij}, \frac{1}{det}]$, extending ϕ on $\overline{k}$, is given (in matrix notation) by $(\phi X_{ij}) = A(X_{ij})$. For every maximal ideal M of D, $\phi(M)$ is also a maximal ideal. The maximal ideal M has the form $(X_{11} - b_{11}, X_{12} - b_{12}, \ldots, X_{dd} - b_{dd})$ and corresponds to the matrix $B = (b_{ij}) \in Gl(d)(\overline{k})$. A small calculation shows that the maximal ideal $\phi(M)$ corresponds to the matrix $A^{-1}\phi(B)$. The expression $\phi(B)$ for a matrix $B = (b_{ij})$ is defined as before as $(\phi(b_{ij}))$. Thus ϕ on D induces the map τ on $Gl(d)(\overline{k})$, given by the formula $\tau(B) = A^{-1}\phi(B)$. The elements $f \in D$ are seen as functions on $Gl(d)(\overline{k})$. The following formula holds

$$(\phi f)(\tau(B)) = \phi(f(B)) \text{ for } f \in D \text{ and } B \in Gl(d)(\overline{k}).$$

Indeed, one can easily verify the formula for $f \in \overline{k}$ and for the $f = X_{ij}$. This proves the formula for any $f \in D$.

For an ideal $J \subset k[X_{ij}, \frac{1}{det}]$ satisfying $\phi(J) \subset J$, one has $\phi(J) = J$. Indeed, if $\phi(J)$ is a proper subset of J then one finds an infinite chain of ideals $J \subset \phi^{-1}(J) \subset \phi^{-2}(J) \subset \ldots$ This contradicts the Noetherian property of $k[X_{ij}, \frac{1}{det}]$. Likewise for reduced algebraic subsets Z of $Gl(d)_k$ the condition $\tau(Z) \subset Z$ implies $\tau(Z) = Z$. The following lemma is an immediate consequence of the remarks above and the formula.

Lemma 1.10 *The ideal J of a reduced subset Z of $Gl(d)_k$ satisfies $\phi(J) = J$ if and only if $Z(\overline{k})$ satisfies $\tau Z(\overline{k}) = Z(\overline{k})$*

An ideal I maximal among the ϕ-invariant ideals corresponds then to a minimal (reduced) algebraic subset Z of $Gl(d)_k$ such that $\tau(Z(\overline{k})) = Z(\overline{k})$. We shall call such a set *a minimal $\tau-invariant$ reduced set.*

Let Z be a minimal τ-invariant reduced subset of $Gl(d)_k$ with ideal $I \subset k[X_{i,j}, \frac{1}{det}]$ and let $O(Z) = k[X_{i,j}, \frac{1}{det}]/I$. Let $x_{i,j}$ denote the image of $X_{i,j}$ in $O(Z)$. One considers the rings

$$k[X_{i,j}, \frac{1}{det}] \subset O(Z) \otimes_k k[X_{i,j}, \frac{1}{det(X_{i,j})}] =$$

$$O(Z) \otimes_C C[Y_{i,j}, \frac{1}{det(Y_{i,j})}] \supset C[Y_{i,j}, \frac{1}{det(Y_{i,j})}] \qquad (1.2)$$

where the variables $Y_{i,j}$ are defined by $(X_{i,j}) = (x_{i,j})(Y_{i,j})$. Note that the action of ϕ on $C[Y_{i,j}, \frac{1}{det(Y_{i,j})}] \subset O(Z) \otimes_k k[X_{i,j}, \frac{1}{det(X_{i,j})}]$ is the identity. Let (I) be the ideal of $O(Z) \otimes_k k[X_{i,j}, \frac{1}{det}]$ generated by I and let J be the intersection of (I) with $C[Y_{i,j}, \frac{1}{det}]$. The ideal (I) is ϕ-invariant. Using that the set of constants of $O(Z)$ is C one can prove that J generates the ideal (I) in $O(Z) \otimes_k k[X_{i,j}, \frac{1}{det}]$. The proof follows from the next lemma.

Lemma 1.11 *Let R be a Picard-Vessiot ring over a field k and let A be a commutative algebra with unit over C_k. The action of ϕ on A is supposed to be the identity. Let N be an ideal of $R \otimes_C A$ which is invariant under ϕ. Then N is generated by the ideal $N \cap A$ of A.*

Proof: Dividing A by $N \cap A$ and $R \otimes_C A$ by the ideal generated by $N \cap A$, one reduces the lemma to proving that $N \neq 0$ implies that $N \cap A \neq 0$. Let $\{a_i\}_{i \in \mathcal{I}}$ be a basis of A over C. Consider a minimal subset $\mathcal{J}$ of $\mathcal{I}$ such that $N \cap \sum_{i \in \mathcal{J}} R \otimes a_i \neq 0$. Fix some $j \in \mathcal{J}$, then the set of the $b \in R$ such that there exists an element in $N \cap \sum_{i \in \mathcal{J}} R \otimes a_i$ with coordinate b at the place j, is a nonzero ideal of R which is invariant under ϕ. Hence there exists an element $f \in N \cap \sum_{i \in \mathcal{J}} R \otimes a_i$ with coordinate 1 at the place j. If $\mathcal{J}$ has only one element then $a_j \in N \cap A$. If $\mathcal{J}$ has more than one element, then $\phi(f) - f$ has a smaller support than $\mathcal{J}$. Hence $\phi(f) - f = 0$. It follows that all the coordinates of f are in the field of constants C of R. Hence $f \in N \cap A$. ∎

In particular, the above lemma shows that when we divide the rings in the sequence (1.2) by the ideals $I, (I)$, and J, we have

$$O(Z) \to O(Z) \otimes_k O(Z) = \qquad (1.3)$$

$$= O(Z) \otimes_C (C[Y_{i,j}, \frac{1}{det(Y_{i,j})}]/J) \leftarrow C[Y_{i,j}, \frac{1}{det(Y_{i,j})}]/J$$

We now assume that the ring $O(Z)$ is a *separable extension* of k ([15], §7, n°. 5). In this case Corollaire 3 of ([15], §7, n°. 5) and Corollaire 3 of ([15], §7, n°. 6) imply that $O(Z) \otimes_k O(Z)$ is reduced. Therefore,

$C[Y_{i,j}, \frac{1}{det(Y_{i,j})}]/J$ is reduced and so J is a radical ideal. Note that our assumption on $O(Z)$ is always true if the characteristic of k is zero or more generally, if k is perfect. We will now show that J is the ideal of an algebraic subgroup of $Gl(d)(C)$.

Consider a matrix $A \in Gl(d)(C)$. Let σ_A denote the action on the three rings in the sequence (1.2) given by $(\sigma_A X_{i,j}) = (X_{i,j})A$ and $(\sigma_A Y_{i,j}) = (Y_{i,j})A$. Using Lemma 1.11 and the facts that I is maximal and Z is minimal, one can easily show that following conditions on A are equivalent:

1. $ZA = Z$.

2. $ZA \cap Z \neq \emptyset$.

3. $\sigma_A I = I$.

4. $I + \sigma I$ is not the unit ideal of $k[X_{i,j}, \frac{1}{det}]$.

5. $\sigma_A(I) = (I)$.

6. $(I) + \sigma_A(I)$ is not the unit ideal of $O(Z) \otimes_k k[X_{i,j}, \frac{1}{det}]$.

7. $\sigma_A J = J$.

8. $J + \sigma_A J$ is not the unit ideal of $C[Y_{i,j}, \frac{1}{det}]$.

The collection of the A satisfying the equivalent conditions form a group.

Lemma 1.12 *Let $O(Z)$ be a separable extension of k. Using the above notation, A satisfies the equivalent conditions if and only if A lies in the reduced subspace V of $Gl(d)_C$ defined by J. Therefore, the set of such A is an algebraic group.*

Proof: Assume that A satisfies the conditions. Condition 3. implies that A defines a difference automorphism on $O(Z)$. We again refer to this automorphism as σ_A. This in turn allows us to define a difference homomorphism $id \otimes \sigma_A :$ $O(Z) \otimes_k O(Z) \to O(Z)$ given by $a \otimes b \mapsto a\sigma_A(b)$. Restricting this map to $C[Y_{i,j}, \frac{1}{det}]/J \subset O(Z) \otimes_k O(Z)$ we get a difference map from $C[Y_{i,j}, \frac{1}{det}]/J$ to $O(Z)$. Since the difference operator is the identity on $C[Y_{i,j}, \frac{1}{det}]/J$, the image of this map must lie in the constants of $O(Z)$, that is, in C. Therefore A corresponds to a map in $HOM_C(C[Y_{i,j}, \frac{1}{det}]/J, C)$ and so A is a point of V.

Conversely, let A lie in V. Then A yields a difference homomorphism from $O(V) \otimes_C C[Y_{i,j}, \frac{1}{det}]/J$ to $O(V)$ given by $a \otimes b \mapsto a \cdot b(A)$. If we restrict this map to $O(Z) = 1 \otimes O(Z) \subset O(Z) \otimes_k O(Z) = O(V) \otimes_C C[Y_{i,j}, \frac{1}{det}]/J$, we have a difference homomorphism from $O(Z)$ to $O(Z)$. One then sees that this yields $\sigma_A I = I$.   ∎

Let G denote the group of the automorphisms of $O(Z)$ over k which commute with the action of ϕ. The group G is called the *(difference) Galois group* of the equation $\phi(Y) = AY$ over the field k. Each element σ of G must have the form $(\sigma x_{i,j}) = (x_{i,j})A$ where $A \in Gl(d)(C)$ is such that σ_A (as defined above) satisfies $\sigma_A I = I$. It follows that G coincides the points of the algebraic group V. In the sequel we will identify G and V and denote by $O(G)$ the ring $C[Y_{i,j}, \frac{1}{det}]/J$. Let $O(G_k) = O(G) \otimes_C k$ and $G_k = spec(O(G_k))$. From the sequence of rings (1.3), we have

$$O(Z) \to O(Z) \otimes_k O(Z) = O(Z) \otimes_C O(G) = O(Z) \otimes_k O(G_k) \qquad (1.4)$$

The first embedding of rings corresponds to the morphism $Z \times G_k \to Z$ given by $(z, g) \mapsto zg$. The identification $O(Z) \otimes_k O(Z) = O(Z) \otimes_C O(G) = O(Z) \otimes_k O(G_k)$ corresponds to the fact that the morphism $Z \times G_k \to Z \times Z$ given by $(z, g) \mapsto (zg, z)$ is an isomorphism. In other words, Z is a k-homogeneous space for G_k or in the language of [20], Z/k is a G-torsor. The following theorem summarizes the above.

Theorem 1.13 *Let R be a separable Picard-Vessiot ring over k, a difference field with algebraically closed subfield of constants, and let G denote the group of the k-algebra automorphisms of R which commute with ϕ. Then G has a natural structure as reduced linear algebraic group over C and the affine scheme $Z = spec(B)$ over k has the structure of a G-torsor over k.*

Example 1.14 In the course of the proof of the above result, the assumption that R was separable over k was used to prove that the group G was a *reduced* space. We give here an example in characteristic p where this is not the case. Let k_0 be an algebraically closed field of characteristic $p > 0$ such that there is an $\alpha \in k_0^*$ which is not of finite order. Let $k = k_0((z))$ with automorphism ϕ given by $\phi(z) = \alpha z$ (and so $\phi(\sum a_n z^n) = \sum a_n \alpha^n z^n$). Let $\beta \in k_0$ satisfy $\beta^p = \alpha$. Consider the 1-dimensional difference equation $\phi(X) = \beta X$. A calculation shows that $L = k[X]/(X^p - z)$ is the Picard-Vessiot ring for the equation. It is in fact an inseparable extension of k and so $L \otimes_k L$ has nilpotents. Following the above development further, one finds that the difference Galois group for the equation above is the group μ_p in characteristic p. This group is given as $spec(k_0[t]/(t^p - 1))$ and the group structure is given by $t \mapsto t \otimes t$.   ∎

The fact that a Picard-Vessiot ring is the coordinate ring of a torsor for its Galois group has several interesting consequences, which we now state.

Corollary 1.15 *Let R be a separable Picard-Vessiot ring over k, a difference field with algebraically closed subfield of constants, and let G denote the group of the k-algebra automorphisms of R which commute with ϕ. The set of G-invariant elements of R is k and R has no proper, nontrivial G-invariant ideals.*

Proof: Let $R = O(Z)$ for some G-torsor Z and let $\bar{k}$ be the algebraic closure of k. Any G-invariant element of R defines a regular function on $Z(\bar{k})$. Since

$G(\overline{k})$ acts transitively on $Z(\overline{k})$, any such element must be constant. The zeroes of an invariant ideal form an invariant subset of $Z(\overline{k})$. Since the action of G is transitive, such a set is empty or the whole space.   ∎

Corollary 1.16 *Let R be the Picard-Vessiot ring over k, a difference field with algebraically closed subfield of constants. There exist idempotents $e_0, \ldots, e_{t-1} \in R$ such that*

1. *$R = R_0 \oplus \cdots \oplus R_{t-1}$ where $R_i = e_i R$,*

2. *$\phi(e_i) = e_{i+1} \ (mod \ t)$ and so ϕ maps R_i isomorphically onto $R_{i+1} (mod \ t)$ and ϕ^t leaves each R_i invariant.*

3. *For each i, R_i is a domain and is a Picard-Vessiot extension of $e_i k$ with respect to ϕ^t.*

Proof: We will present two proofs. The first uses the Galois group G and the fact that $R = O(Z)$ for some G-torsor Z. It is therefore only valid under the additional assumption that R is separable over k. The second goes back to first principals and does not require this additional assumption.

Proof 1: Let $R = O(Z)$ for some G-torsor Z and let $\overline{k}$ be the algebraic closure of k. Since $G(\overline{k})$ acts transitively on $Z(\overline{k})$, this latter algebraic set must be smooth. Therefore, the k-irreducible components $Z_0, \ldots Z_{t-1}$ of Z must be disjoint and so must be the irreducible components of Z. Therefore $O(Z)$ is equal to the product of the rings $R_i = O(Z_i)$. This gives conclusion 1. above. Since Z is a minimal τ-invariant set, τ must act as a cyclic permutation on the components of Z. Renumbering, if necessary we can assume that $\tau(Z_i) = Z_{i+1}(mod \ t)$ and ϕ^t leaves each R_i invariant. This gives conclusion 2. To prove conclusion 3. we must show that each Z_i is a minimal τ^t-invariant set. Let W be a proper τ^t invariant subset of Z_0. We then have that $Y = \cup_{i=0}^{t-1} \tau(W)$ is a τ invariant subset of Z. Since $Y(\overline{k}) \cap Z_0(\overline{k}) \neq \emptyset$, we have that $Z_0(\overline{k}) \subset Y(\overline{k})$. Since $Z_0(\overline{k})$ is k-irreducible and $Z_0(\overline{k}) \cap W(\overline{k}) \neq \emptyset$, we must have $Z_0(\overline{k}) \subset W$. Therefore $Z_0 = W$ and so Z_0 is a minimal τ^t-invariant set.

Proof 2: Since R has no nilpotent elements and is finitely generated over k, we may write $(0) = \cap_{i=0}^{t-1} I_i$, where the I_i are prime ideals. We shall assume that this is a minimal representation and so is unique (Theorems 4 and 6 of Ch. IV, Sec. 5 of [63]). Since $(0) = \cap_{i=0}^{t-1} \sigma(I_i)$ as well, we see that σ leaves the set of I_i invariant. For each i, $\cap_{j>0} \sigma^j(I_i)$ is a difference ideal and so must be (0). Using the uniqueness of the minimal representation, we have that $\{I_1, \sigma(I_1), \ldots \sigma^{t-1}(I_1)\} = \{I_0, I_2, \ldots, I_{t-1}\}$, so after a possible renumbering, we can assume that $\sigma(I_i) = I_{i+1} \ (mod \ t)$.

For each i, let $J_i = \{r \in R \mid \sigma^t(r) \in I_i\}$. For each i, J_i is a prime ideal containing I_i. Furthermore, $\cap_{i=0}^{t-1} J_i$ is a proper difference ideal and so must be (0). Therefore, for each i, $J_i = I_i$, so $r \in I_i$ if and only if $\sigma^t(r) \in I_i$. This implies that the ring $S_i = R/I_i$ is a difference ring with respect to σ^t. Let $\pi_i : R \to S_i$ be the canonical homomorphism. Note that σ induces an isomorphism $\sigma_i : S_i \to S_{i+1}$.

We now claim that for each i, S_i has no non-zero proper σ^t–invariant ideals. It is enough to show that R has no proper σ^t–invariant ideals properly containing I_i. Assume the contrary and let J_i be such an ideal. One then sees that $\cap_{j=0}^{t-1} \sigma^j(J_i)$ is a proper difference ideal and so must be (0). In particular, we have that $\cap_{t=0}^{t-1} \sigma^t(J_i) \subset I_i$. Since the I_j are prime we must have that for some $t_0 \leq t - 1$, $\sigma^{t_0}(I_i) \subset \sigma^{t_0}(J_i) \subset I_i$. This implies that $I_{i+t_0 \ (mod \ t)} = I_i$, a contradiction unless $t_0 = 0$, in which case we have $J_i = I_i$. Therefore each S_i is a simple difference ring with respect to σ^t. S_i is generated (over $\pi_i(k)$) by the entries of $\pi_i(Y)$ and $\det(\pi_i(Y)^{-1}$. Note that $\sigma^t(\pi(Y)) = B_i \pi(Y)$, for some $B_i \in Gl_n(\pi_i(k))$ so $\pi(Y)$ is a fundamental matrix. Therefore S_i is a Picard-Vessiot extension of $\pi_i(k)$ with respect to σ^t.

We further claim that the I_i are pairwise comaximal, i.e., $I_i + I_j = R$ for $i \neq j$. To see this note that $I_i + I_j$ is a σ^t–invariant ideal of R containing both I_i and I_j. By the above, it must be all of R. Theorem 31 of ([63], Ch. III, Sec. 13) implies that the map $\pi : R \to \oplus_{i=0}^{t-1} S_i$ given by $\pi(r) = (\pi_0(r), \ldots, \pi_{t-1}(r))$ is an isomorphism of rings. The ring $\oplus_{i=0}^{t-1} S_i$ has the structure of a difference ring where the automorphism $\tilde{\sigma}(r_0, \ldots, r_{t-1}) = (\sigma_{t-1}(r_{t-1}), \sigma_0(r_1), \ldots, \sigma_{t-2}(r_{t-2}))$. With this structure, π is a k–isomorphism of difference rings. Letting $R_i = \pi^{-1}(S_i)$ and $e_i = \pi^{-1}(id_i)$, where id_i is the identity in S_i, we achieve the conclusions above. ∎

Let k be a difference field with algebraically closed constants. Let R be a Picard-Vessiot extension of k and let $R = R_0 \oplus \cdots \oplus R_{t-1}$, $R_i = e_i R$ be as in Proposition 1.16. Recall that $\sigma : R_i \to R_{i+1}$ is an isomorphism and R_0 is a Picard-Vessiot extension of k with respect to the automorphism σ^t. We will define a map $\Gamma : Gal(R_0/k) \to Gal(R/k)$. Let $\psi \in Gal(R_0/k)$. Define $\Gamma(\psi) = \phi$ where for $r = (r_0, \ldots, r_{t-1}) \in R$, $\phi(r_0, \ldots, r_{t-1})$,

$$\phi(r_0, \ldots, r_{t-1}) = (\psi(r_0), \sigma \psi \sigma^{-1}(r_1), \sigma^2 \psi \sigma^{-2}(r_2), \ldots, \sigma^{t-1} \psi \sigma^{1-t}(r_{t-1}))$$

We also define a map $\Delta : Gal(R/k) \to \mathbf{Z}/t\mathbf{Z}$. Given $\phi \in Gal(R/k)$, one sees that ϕ permutes the e_i. Let $\phi(e_0) = e_j$. Define $\Delta(\phi) = j$

Corollary 1.17 *Let R be a separable Picard-Vessiot extension over k, a difference field with algebraically closed subfield of constants and let the maps Γ and Δ be defined as above. If $\psi \in Gal(R_0, k)$ then $\Gamma(\psi) \in Gal(R/k)$ and we have the following exact sequence*

$$0 \longrightarrow Gal(R_0/k) \xrightarrow{\Gamma} Gal(R/k) \xrightarrow{\Delta} \mathbf{Z}/t\mathbf{Z} \longrightarrow 0$$

Proof: One can easily verify that $\Gamma(\psi)$ is a k−algebra isomorphism. To see that it commutes with σ one computes

$$\Gamma(\psi)(\sigma((r_0,\ldots,r_{t-1}))) = \Gamma(\psi)((\sigma(r_{t-1}),\sigma(r_0),\ldots,\sigma(r_{t-2}))$$

$$= (\psi(\sigma(r_{t-1})),\sigma\psi\sigma^{-1}(\sigma(r_0)),\sigma^2\psi\sigma^{-2}(\sigma(r_1)),\ldots,\sigma^{t-1}\phi\sigma^{1-t}(\sigma(r_{t-2}))$$

$$= (\psi\sigma(r_{t-1}),\sigma\psi(r_0),\sigma^2\psi\sigma^{-1}(r_1),\ldots,\sigma^{t-1}\phi\sigma^{2-t}(r_{t-2}))$$

$$= (\sigma^t\psi\sigma^{1-t}(r_{t-1}),\sigma\psi(r_0),\sigma^2\psi\sigma^{-1}(r_1),\ldots,\sigma^{t-1}\phi\sigma^{2-t}(r_{t-2}))$$

$$= \sigma(\psi(r_0),\sigma\psi\sigma^{-1}(r_1),\ldots,\sigma^{t-1}\phi\sigma^{1-t}(r_{t-1})))$$

$$= \sigma(\Gamma(\psi)((r_0,\ldots,r_{t-1})))$$

Notice that the equality $\psi\sigma(r_t) = \sigma^t\psi\sigma^{1-t}(r_t)$ needed to go from the third to the fourth line follows from the fact that ψ commutes with σ^t. One can readily verify that Γ is an injective homomorphism.

We now turn to Δ. If $\phi(e_0) = e_i$, then $\phi(e_j) = \phi(\sigma^j(e_0)) = \sigma^j(\phi(e_0)) = e_{i+j}(mod\ t)$. From this one can readily see that Δ is a homomorphism of $Gal(R/k)$ into $\mathbf{Z}/t\mathbf{Z}$. If $\Delta(\phi) = 0$, the ϕ leaves R_0 invariant. One easily sees that $\phi\mid_{R_0}\in Gal(R_0/k)$ and that $\phi = \Gamma(\phi\mid_{R_0})$. Therefore, $Ker(\Delta) = Im(\Gamma)$. Finally we need to show that Δ is surjective. Let $j \in \mathbf{Z}/t\mathbf{Z}$ generate the image of Δ. Note that j divides t. Let $f = e_0 + e_j + e_{2j} + \ldots + e_{(t-1)j}$. One sees that $\phi(f) = f$ for all $\phi \in Gal(R/k)$ and so Corollary 1.15 implies that $f \in k$. Since $f^2 = f$, we have that $f = 0$ or $f = 1$. Clearly $f \neq 0$ so $f = 1$ which can only happen if $j = 1$. Therefore, Δ is surjective.   ∎

The fact that the Z/k is a G−torsor implies that for any field $L \supset k$ such that $Z(L) \neq \emptyset$ and for any $B \in Z(L)$, $Z(L) = B \cdot G(L)$. This is certainly the case if $L = \bar{k}$. For any $B \in Z(\bar{k})$ and $\sigma \in Gal(\bar{k}/k)$ the element $B^{-1}\sigma(B) \in G(\bar{k})$ and the map $\sigma \mapsto B^{-1}\sigma(B)$ is a 1−cocycle in $H^1(Gal(\bar{k}/k),G(\bar{k}))$ (c.f., [57]). This cocycle is independent of the choice of $B \in Z(\bar{k})$. In particular if $H^1(Gal(\bar{k}/k),G(\bar{k})) = 0$ then all G−torsors over k are isomorphic to G_k endowed with the right regular action of G. We summarize this observation in the following corollary.

Corollary 1.18 *Let R be a separable Picard-Vessiot ring over k, a difference field with algebraically closed subfield of constants, and let G be the Galois group. If $H^1(Gal(\bar{k}/k),G(\bar{k})) = 0$, then $Z = spec(R)$ is G−isomorphic to the G−torsor G_k and so $R = C[G] \otimes k$.*

We note that for any C^1-field k and connected group G, we have $H^1(Gal(\bar{k}/k),G(\bar{k})) = 0$ (c.f., [57], Chapter 3). Each of the fields $C(z)$, $\mathbf{C}(\{z^{-1}\})$,

$C((z^{-1}))$ is a C^1-field. We shall now apply this result to describe the Picard-Vessiot extensions of $C(z)$, C an algebraically closed field of characteristic zero with ϕ_a defined by $\phi_a = z + a$, $a \neq 0$.

Lemma 1.19 ϕ_a *does not extend to any proper finite extension L of $k = C(z)$.*

Proof: Let $u : X \to CP^1$ denote the morphism of smooth algebraic curves corresponding to $k \subset L$. If ϕ_a extends to L, then ϕ_a will permute the finite set of ramification points of u. Since the only finite set in CP^1 left invariant by ϕ_a is $\{\infty\}$, the only possible ramification point of u is ∞. This implies that the degree of u is 1 so $L = k$. ∎

Proposition 1.20 *Let R be a Picard-Vessiot extension of $k = C(z)$ with Galois group G. The corresponding $G-$torsor Z has a point which is rational over k and so Z and G_k are isomorphic. Moreover, G/G^0 is cyclic.*

Proof: Let $Z_0, \ldots, Z_{t-1}$ be the $k-$components of Z. In the proof of Corollary 1.16 we saw that $R = O(Z) = O(Z_0) \mp \ldots \oplus O(Z_{t-1})$. Lemma 1.19 implies that for each i, k is relatively algebraically closed in $O(Z_i)$. Therefore, $O(Z_i) \otimes_k \overline{k}$ is a domain and the Z_i remain irreducible over $\overline{k}$. Since $Z(\overline{k}) = B \cdot G(\overline{k})$, we see that some component, say Z_0 is a G^0-torsor. As noted above, k is a C^1 field, so Corollary 1.18 implies that $Z_0(k)$ is nonempty. Fix some $B \in Z_0(k)$. One then has that $Z_0 = BG_k^0$ and $Z = BG_k$. Then $BG_k = \tau(BG_k) = A^{-1}\phi(B)G_k$, where R is a Picard-Vessiot extension for the system $\phi(Y) = AY$. We therefore have $B^{-1}A^{-1}\phi(B) \in G_k$. Choose $N \in G(C)$ such that $B^{-1}A^{-1}\phi(B) \in G^0(k)N$. Let H be the group generated by G^0 and N. Then $\tau(BH_k) = A^{-1}\phi(B)H_k = B(B^{-1}A^{-1}\phi(B))H_k = BH_k$ since $B^{-1}A^{-1}\phi(B) \in G^0(k)N \subset H_k(k)$. The minimality of Z implies that $H = G$ and hence that G/G_0 is cyclic. ∎

The above result gives us a characterization of the Galois group of a Picard-Vessiot extension of $C(z)$.

Proposition 1.21 *Let $k = C(z)$ and G an algebraic subgroup of $Gl(d)_C$. Let $\phi(Y) = AY$ be a difference system with $A \in G(k)$. Then*

1. *the Galois group of $\phi(Y) = AY$ over k is a subgroup of G_C.*

2. *any minimal element in the set of $C-$subgroups H of G for which there exists a $B \in Gl(d)(k)$ with $B^{-1}A^{-1}\phi(B) \in H(k)$ is the Galois group of $\phi(Y) = AY$ over k.*

3. *the Galois group of $\phi(Y) = AY$ over k is G if and only if for any $B \in G(k)$ and any proper $C-$subgroup H of G one has that $B^{-1}A^{-1}\phi(B) \notin H(k)$.*

Proof: 1. Note that $G(\overline{k})$ is a τ-invariant subset of $Gl(d)(\overline{k})$. Therefore, there exists a minimal τ-invariant subset Z of G_k. The above Proposition shows that $Z = BH_k$ where the C-group H is the Galois group and $B \in Gl(d)(k)$. This implies that $B \in G_k(k)$ and so $H(k) \subset G(k)$. Therefore $H \subset G$.

2. For any C-group H and $B \in Gl(d)(k)$, BH_k is τ-invariant if and only if $B^{-1}A^{-1}\phi(B) \in H(k)$. Any minimal τ-invariant subset of BH_k is of the form $\overline{B}\,\overline{H}_k$ for some C-subgroup $\overline{H} \subset H$. Therefore any minimal element of the above set of subgroups corresponds to a minimal τ-invariant set and so must be the Galois group.

3. This follows from 2. ∎

1.3 Galois correspondence for difference equations

One cannot expect that the Picard-Vessiot rings for difference equations, that we have introduced, are the correct objects for the Galois correspondence. Indeed, such a Picard-Vessiot ring could be $R = k \otimes_C C[G]$ where k is a difference field with algebraically closed field of constants C and where $C[G]$ is the ring of regular functions on an algebraic group G defined over C. For an algebraic subgroup H of G the ring of invariants R^H is the ring of regular functions on $(G/H)_k$. In some cases, e.g. $G = Gl(n, C)$ and H is a Borel subgroup, the space G/H is a connected projective variety and so the ring of regular functions on $(G/H)_k$ is just k. The same example could be used for the case of differential equations. For differential equations one considers differential fields instead of differential rings. For those differential fields the Galois correspondence holds.

For difference equations this method cannot be used since there is in general no difference field attached to a difference equation over k. Yet we can associate the following with a difference equation.

Definition 1.22 *Let k be difference field and K a difference ring with $k \subset K$. Let $A \in Gl(n)(k)$. We say that K is the total Picard-Vessiot ring K of the equation $\phi(Y) = AY$ if K is the total ring of fractions of the Picard-Vessiot ring R of the equation.*

A Picard-Vessiot ring R is a product of domains $R = R_0 \oplus \ldots \oplus R_{t-1}$. Each component R_i is invariant under the action of ϕ^t. The automorphism ϕ of R permutes the factors of this product in a cyclic way. The total Picard-Vessiot ring K is therefore a product of fields $K = K_0 \oplus \ldots \oplus K_{t-1}$. Each K_i is the field of fractions of R_i. The fields K_i are invariant under ϕ^t. The map ϕ permutes the factors in a cyclic way. More precisely, $\phi(K_i) = K_{i+1}$ (with a cyclic notation).

The following gives an alternate description of the total Picard-Vessiot ring. Note that we are restricting ourselves to perfect fields. This is to insure that Picard-Vessiot extensions are separable and that the theory described in section 1.2 applies.

Proposition 1.23 *Let k be a perfect field and let $\phi(Y) = BY$ be a difference equation over the difference field k with an algebraically closed field of constants C. Let the difference ring extension $L \supset k$ have the following properties:*

1. *L has no nilpotent elements and every non-zero divisor of L is invertible.*

2. *The set of constants of L is C.*

3. *There is a fundamental matrix F for the equation with entries in L.*

4. *L is minimal with respect to 1., 2. and 3..*

Then L is k-isomorphic as a difference ring with the total Picard-Vessiot ring of the equation.

Proof: Let $(X_{i,j})$ denote a matrix of variables. One considers as before the difference ring $k[X_{i,j}, \frac{1}{\det}]$, with the action of ϕ given by the formula $(\phi X_{i,j}) = B(X_{i,j})$. From 1., 3. and 4. it follows that L is the total ring of fractions of $k[X_{i,j}, \frac{1}{\det}]/I$, where I is some radical, ϕ-invariant ideal. If we can show that condition 2. implies that I is a maximal ϕ-invariant ideal in $k[X_{i,j}, \frac{1}{\det}]$, then $k[X_{i,j}, \frac{1}{\det}]/I$ is a Picard-Vessiot extension for the equation and L is isomorphic to the total Picard-Vessiot ring of the equation.

Let R denote a Picard-Vessiot extension for the equation and let G denote the difference Galois group of the equation. Consider any k-algebra A (commutative and with a unit element). The action of G on A is supposed to be trivial. On $A \otimes_k R$ the action of any $\sigma \in G$ is given by $\sigma(a \otimes r) = a \otimes (\sigma r)$. The map $A \to A \otimes_k R$ has the property:

(*) *The map $I \mapsto (I) = IA \otimes_k R$ from the set of ideals of A to the set of G-invariant ideals of $A \otimes_k R$ is bijective.*

The proof of this statement is similar to the proof of Lemma 1.11. To modify this latter proof in order to establish (*) one needs to use the facts that $R^G = k$ and that the only G-invariant ideals of R are $\{0\}$ and R. These facts were established in Corollary 1.15.

As before, we consider the sequence of rings

$$k[X_{i,j}, \frac{1}{\det}] \subset R \otimes_k k[X_{i,j}, \frac{1}{\det}] = R \otimes_C C[Y_{i,j}, \frac{1}{\det}] \supset C[Y_{i,j}, \frac{1}{\det}],$$

where the variables $Y_{i,j}$ are defined by $(X_{i,j}) = (x_{i,j})(Y_{i,j})$ with $(x_{i,j})$ a fundamental matrix for the equation with coordinates in R. The action of ϕ on the variables $Y_{i,j}$ is the identity. The action of G on the variables $Y_{i,j}$ is determined by the assumption that G acts as the identity on the variables $X_{i,j}$. We claim now that there is a bijection between the set of ϕ-invariant ideals of $k[X_{i,j}, \frac{1}{det}]$ and the set of G-invariant ideals of $C[Y_{i,j}\frac{1}{det}]$. This bijection is given by

$$I \mapsto J := (IR \otimes_k k[X_{i,j}, \frac{1}{det}]) \cap C[Y_{i,j}, \frac{1}{det}].$$

The claim follows from statement (*) and Lemma 1.11. Since k is perfect, if I is a radical ideal then J is also a radical ideal.

Let I be again the radical ϕ-invariant ideal such that L is the total ring of fractions of $k[X_{i,j}, \frac{1}{det}]/I$. We suppose now that I is not a maximal ϕ-invariant ideal. The corresponding ideal J is a radical G-invariant ideal but not a maximal G-invariant ideal. The algebraic subspace W of $Gl(n, C)$ defined by the ideal J is invariant under the action of G, since J is G-invariant. The quotient space $W/G \subset Gl(n, C)/G$ is not one point and therefore there exists a non-constant invertible f in the total ring of quotients of $C[Y_{i,j}, \frac{1}{det}]/J$, which is G-invariant. The invertible element $F := 1 \otimes f$ in the total ring of fractions of $R \otimes_C C[Y_{i,j}, \frac{1}{det}]/J = R \otimes_k k[X_{i,j}, \frac{1}{det}]/I$ is G-invariant and also ϕ-invariant. Let K denote the total ring of fractions of R and L the total ring of fractions of $k[X_{i,j}, \frac{1}{det}]/I$. Both K and L are finite products of field extensions of k. Their tensor product $K \otimes_k L$ is again a finite product of field extensions of k (since k is perfect). Hence $K \otimes_k L$ is the total ring of fractions of $R \otimes_k k[X_{i,j}, \frac{1}{det}]/I$. The element $F \in K \otimes_k L$ is G-invariant. Using the fact that $K^G = k$ one finds that $F = 1 \otimes h$ with $h \in L$. This element h is not in C and is ϕ-invariant. This contradicts property 2. of L.   ∎

The following Corollary is an immediate consequence of the above proposition.

Corollary 1.24 *Let k be a perfect field and let $\phi(Y) = AY$ be a difference equation over the difference field k with algebraically closed field of constants C. Let the difference ring $R \supset k$ have the following properties:*

1. *R has no nilpotent elements*

2. *The set of constants of the total quotient ring of R is C*

3. *There is a fundamental matrix F for the equation with entries in R.*

4. *R is minimal with respect to 1., 2., and 3.*

Then R is a Picard-Vessiot ring of the equation.

The following example shows that, in the above corollary, condition 2. cannot be replaced by the weaker condition *The set of constants of R is C*.

Example 1.25 Let k be a difference field with an algebraically closed field of constants C_k. Let $a, b, d \in k$, $ad \in k^*$ be chosen so that the equation

$$\phi(Y) = \begin{pmatrix} a & b \\ 0 & d \end{pmatrix} Y$$

has differential Galois group

$$B = \{ \begin{pmatrix} \alpha & \beta \\ 0 & \delta \end{pmatrix} \mid \alpha, \beta, \delta \in C, \alpha\delta \neq 0 \}.$$

In Chapter 3 we show that this can always be done for $k = C(z), \phi(z) = z + 1$. Let R be the Picard-Vessiot ring for this equation. This ring has an obvious ϕ-action and B-action. Let $X_{1,1}, X_{1,2}, X_{2,1}, X_{2,2}$ be indeterminates and let $A = k[X_{1,1}, X_{1,2}, X_{2,1}, X_{2,2}, \frac{1}{det}]$ with ϕ-action given by $(\phi X_{i,j}) = \begin{pmatrix} a & b \\ 0 & d \end{pmatrix} (X_{i,j})$ and trivial B-action. Consider the sequence of rings

$$A \subset R \otimes_k A = R \otimes_C C[Y_{i,j}, \frac{1}{det}] \supset C[Y_{i,j}, \frac{1}{det}],$$

where the variables $Y_{i,j}$ are defined by $(X_{i,j}) = (x_{i,j})(Y_{i,j})$ with $(x_{i,j})$ a fundamental matrix for the equation with coordinates in R. The action of ϕ on the variables $Y_{i,j}$ is the identity. The action of G on the variables $Y_{i,j}$ is determined by the assumption that G acts as the identity on the variables $X_{i,j}$. Using superscripts to denote the ring of elements fixed by the designated actions, we have

$$A^\phi \subset (R \otimes_k A)^{<B,\phi>} = R \otimes_C C[Y_{i,j}, \frac{1}{det}]^{<B,\phi>} = (C[Y_{i,j}, \frac{1}{det}])^B = C.$$

The last equality follows from the fact that B is a Borel subgroup of $Gl(2)$ and so $Gl(2)/B$ is a projective variety. The ring $(C[Y_{i,j}, \frac{1}{det}])^B$ corresponds to regular functions on this variety and so must be C. Therefore, the set of constants of A is C.

However, the total ring of quotients of A contains $\frac{X_{2,1}}{X_{2,2}}$, which is clearly ϕ-invariant. Therefore, A cannot be a Picard-Vessiot ring. We note that a similar example in the differential case shows that the corresponding weakening of a differential version of the above corollary is not true. ∎

Before giving the statement of the Galois correspondence for total Picard-Vessiot rings we make a further study of those rings. Let $R = R_0 \oplus \dots \oplus R_{t-1}$

be the Picard-Vessiot ring of the equation $\phi(Y) = AY$. We consider now the difference field (k, ϕ^t), i.e. the automorphism ϕ is replaced by ϕ^t, and we consider the difference equation $\phi^t(Y) = A_t Y$ with $A_t = \phi^{t-1}(A) \ldots \phi^2(A)\phi(A)A$.

Lemma 1.26 *Each of the components R_i of R is a Picard-Vessiot ring for the equation $\phi^t(Y) = A_t Y$ over the difference field (k, ϕ^t).*

Proof: Let F be a fundamental matrix with coefficients in R for the equation $\phi(Y) = AY$ over (k, ϕ). Then F is also a fundamental matrix for the equation $\phi^t(Y) = A_t Y$ over (k, ϕ^t). The image F_i of F in $Gl(n, R_i)$ under the map $R \to R_i$ is again a fundamental matrix. The coefficients of F_i generate R_i over (k, ϕ^t). The lemma is proved if we can show that R_i has only trivial ϕ^t-invariant ideals. Let $J_i \subset R_i$ be a non zero ϕ^t-invariant ideal. The action of ϕ on R permutes the factors R_j. We use the cyclic notation $R_j = R_{j+mt}$ for every $m \in \mathbf{Z}$. Define the ϕ^t-invariant ideal $J_j \subset R_j$ by $J_j = \phi^s J_i$ with $s + i \equiv j \mod t$. Then $J_0 \oplus \ldots \oplus J_{t-1}$ is a ϕ-invariant ideal of R and therefore equal to R. Hence $J_i = R_i$. ∎

Corollary 1.27 *Let $d \geq 1$ be a divisor of t. Using a cyclic notation for the indices $\{0, \ldots, t-1\}$ we consider the subrings $\oplus_{m=0}^{(t/d)-1} R_{i+md}$ of $R_0 \oplus \ldots \oplus R_{t-1}$. Each subring is a Picard-Vessiot ring for the equation $\phi^d(Y) = A_d Y$ over the difference field (k, ϕ^d).*

Proof: The proof is similar to that of Lemma 1.26. ∎

The following lemma is a special case of the Galois correspondence given in Theorem 1.29.

Lemma 1.28 *Let K be the total Picard-Vessiot ring of the equation $\phi(Y) = AY$ over the perfect difference field k with algebraically closed field of constants C. Let G denote the difference Galois group of the equation and let H be an algebraic subgroup of G. Then G acts on K and moreover:*

1. *K^G, the set of G-invariant elements of K, is equal to k.*

2. *If $K^H = k$ then $H = G$.*

Proof: 1. The Picard-Vessiot ring is again denoted by R. The ring R is the ring of regular functions on the G-torsor Z over k. Take $f \in K^G$. Then f is seen as a morphism $f : Z \to \mathbf{P}^1_k$, where $\mathbf{P}^1_k$ denotes the projective line over k. Let $pr_i : Z \times_k Z \to Z$, $i = 1, 2$ denote the two projections. The two morphisms

$$Z \times_C G \xrightarrow{\cong} Z \times_k Z \xrightarrow{pr_i} Z \xrightarrow{f} \mathbf{P}^1_k$$

are equal since f is G-invariant. Then also the two morphisms

$$Z \times_k Z \xrightarrow{pr_i} Z \xrightarrow{f} \mathbf{P}^1_k$$

are equal. It follows that f is a constant morphism. In other words $f \in k$.

2. The group $G_k = G \times_C k$ acts on Z and so does the subgroup H_k of G_k. For some finite extension k' of the field k the space $Z_{k'} := Z \times_k k'$ becomes isomorphic to $G_{k'}$. The quotient $Z_{k'}/H_{k'} \cong G_{k'}/H_{k'}$ is an algebraic variety as is well known and therefore the quotient Z/H_k has also the structure of an algebraic variety. The ring of rational functions of Z/H_k coincides with K^H. If $G \neq H$, then Z/H_k is not one point since $G_{k'}/H_{k'}$ is not one point. Therefore K^H is not equal to k. This proves the second statement. ∎

We can now formulate and prove the Galois correspondence for total Picard-Vessiot rings. We restrict ourselves to fields of characteristic zero to avoid considerations of separability.

Theorem 1.29 *Let k be a field of characteristic zero. Let K/k be a total Picard-Vessiot ring over k and let G denote the difference Galois group of the equation. Let $\mathcal{F}$ denote the set of difference rings F with $k \subset F \subset K$ and such that every non zero divisor of F is a unit of F. Let $\mathcal{G}$ denote the set of algebraic subgroups of G.*

1. *For any $F \in \mathcal{F}$ the subgroup $G(K/F) \subset G$ of the elements of G which fix F pointwise, is an algebraic subgroup of G.*

2. *For any algebraic subgroup H of G the ring K^H belongs to $\mathcal{F}$.*

3. *Let $\alpha : \mathcal{F} \to \mathcal{G}$ and $\beta : \mathcal{G} \to \mathcal{F}$ denote the maps, $F \mapsto G(K/F)$ and $H \mapsto K^H$. Then α and β are each other's inverses.*

Proof: The second item is evident.

To see that 1. is true, note that $f \in F$ is a rational function in $x_{i,j}$ with coefficients in k, where $x = (x_{i,j})$ is a fundamental matrix for the equation. Let $f = g/h$ with $g, h \in R$ and let $\sigma \in G(K/k)$. Both g and h are polynomial expressions in the $x_{i,j}$ and $\sigma(x) = x(\sigma_{i,j})$ for some $(\sigma_{i,j}) \in Gl_n(C)$. The equation $g(\sigma(x))h(x) - g(x)h(\sigma(x)) = 0$ is equivalent to a set of polynomial equations in the $\sigma_{i,j}$ with coefficients in C. These equations, for all f in F, define an algebraic subgroup of $G(K/k)$.

Further $F \subset \beta\alpha(F)$ and $H \subset \alpha\beta(H)$ are obvious. We have to show that the two inclusions are in fact equalities. Let $F \in \mathcal{F}$ be given. We have to show that the set of the $G(K/F)$-invariant elements of K coincides with F. A subset A of $\{0, 1, \ldots, t-1\} = \mathbf{Z}/t\mathbf{Z}$ is called a support (for F) if there exists an $f = (f_0, \ldots, f_{t-1}) \in F$ with $f_i \neq 0$ if and only if $i \in A$. If A and B are supports then so are $A \cap B$ and $A \cup B$. Let s denote the shift over 1 on $\mathbf{Z}/t\mathbf{Z}$. If A is a support then $s(A)$ is also a support.

Let A be a minimal support containing 0. Then $s(A), \dots, s^t(A)$ are also minimal supports. If $A \cap s^i(A) \neq \emptyset$ then, by minimality, $A = s^i(A)$. Let d, $1 \leq d \leq t$ be the smallest number with $A = s^d(A)$. The sets $A, s(A), \dots, s^{d-1}(A)$ are disjoint and A contains $0, d, 2d, \dots$. From this one concludes that d is a divisor of t and that $A = \{0, d, \dots, t-d\}$. Choose an element $f \in F$ with support A. Then $g = f + \phi(f) + \dots \phi^{d-1}(f) \in F$ has support $\{0, 1, \dots, t-1\}$. By assumption g is invertible and $E_0 := g^{-1} f = e_0 + e_d + \dots + e_{t-d} \in F$. Put $E_i = \phi^i(E_0)$ for $i = 0, \dots, d-1$. The following steps give the desired result.

- $F = \oplus_{i=0}^{d-1} F E_i$ and each $F E_i$ is a field.

- The fields $F E_i \subset K E_i$ are invariant under ϕ^d.

- According to Corollary 1.27, the ring $K E_i$ is the total Picard-Vessiot ring of the equation $\phi(Y) = A_d Y$ over the difference field (k, ϕ^d). Hence $K E_i$ is also the total Picard-Vessiot field of the same equation over $F E_i$.

- The elements of $G(K/F)$ can be described as the tuples $(\sigma_0, \dots, \sigma_{d-1})$ such that:
  σ_i is an automorphism of $K E_i$ over $F E_i$ commuting with ϕ^d. $\phi \sigma_i = \sigma_{i+1} \phi$ for $i = 0, \dots, d-1$ and with a cyclic notation (modulo d).

- The first part of Lemma 1.28, applied to each $F E_i \subset K E_i$, gives that the set of $G(K/F)$-invariant elements of K is equal to F.

Let $H \in \mathcal{G}$. We have to show that $G(K/K^H)$ is equal to H. If the ring K^H happens to be field, then K is again the total Picard-Vessiot ring of the equation $\phi(Y) = AY$ over K^H and part 2. of Lemma 1.28 finishes the proof.

In the general case we use the description above of any $F \in \mathcal{F}$. This means that there is a divisor $d \geq 1$ of t such that $K^H = K^H E_0 \oplus K^H E_1 \oplus \dots \oplus K^H E_{d-1}$ and each $K^H E_i$ is a field (invariant under ϕ^d). The elements of $G(K/K^H)$ are described as tuples $(\sigma_0, \dots, \sigma_{d-1})$ as before. An application of part 2. of Lemma 1.28 to each of the total Picard-Vessiot extensions $K^H E_i \subset K E_i$ yields that $H = G(K/K^H)$. ∎

When a Picard-Vessiot ring R is a domain, we refer to the total quotient ring as a *Picard-Vessiot field*. In this case a correspondence between *connected* closed subgroups of the Galois group and relatively algebraically closed intermediate fields was proven in [25] and can also be deduced from [8]. The above theorem removes both the hypothesis of R being a domain and the connectedness assumptions.

Corollary 1.30 *The group $H \in \mathcal{G}$ is a normal subgroup if and only if the difference ring $F := K^H$ has the property that for every $z \in F \setminus k$ there is an* *automorphism σ of F/k which commutes with ϕ and satisfies $\sigma z \neq z$. If $H \in \mathcal{G}$ is normal then the group of all automorphisms σ of F/K which commute with ϕ is isomorphic to G/H.*

Proof: A proof following the classical lines works here. ∎

We use the same notation as before. In later sections we will also be interested in a description of the difference ring R^H where the algebraic subgroup H of G contains G^0.

Corollary 1.31 *Suppose that the algebraic group $H \subset G$ contains G^0. The difference ring R^H is a finite dimension vector space over k with dimension equal to $[G : H]$.*

Proof: We shall use the Galois theory of separable algebras developed by Chase, Harrison and Rosenberg in [17]. They consider rings $T \subset S$ and finite groups G of automorphisms of S with $S^G = T$. They define a subring $T \subset U \subset S$ to be G-strong if for any two elements $f, g \in G$ either $f = g$ on U or for any idempotent e of S, there is an element $u \in U$ such that $f(u)e \neq g(u)e$. They show (Theorem 1.3, p. 18 and Theorem 2.2, p. 22) that if S is a separable, G-strong extension of T, then there is the usual Galois correspondence correspondence between separable, G-strong subalgebras of S containing T and subgroups of G. Furthermore, for any subgroup $H \subset G$ one has $\dim_k R^H = [G : H]$.

To apply this theory, we let $S := R^{G^0}$ and $T := k$. The finite group G/G^0 acts on S and its set of invariants is equal to T (by Lemma 1.28). Note that since k has characteristic zero and R is a direct sum of domains, S (and any subalgebra) is a separable k-algebra. We now will show that any difference k-subalgebra U of S is G/G^0-strong. Recall that $R = Re_0 \oplus \ldots \oplus Re_{t-1}$ for some idempotents e_i and that each Re_i is a domain. This implies that any idempotent is of the form $e_{i_1} + \ldots + e_{i_s}$ so it suffices to check the G/G^0−strong condition for each e_i. We will do this for e_0, the other cases being similar. Let $f, g \in G/G^0$ and assume that $f(u)e_0 = g(u)e_0$ for all $u \in U$, a difference k-subalgebra of S. Applying the difference operator σ and recalling that f and g are σ-isomorphisms, we have that $f(\sigma(u))e_1 = g(\sigma(u))e_1$. Since σ is an automorphism of U, we have that $f(u)e_1 = g(u)e_1$ for all $u \in U$. Continuing, one sees that $f(u)e_i = g(u)e_i$ for all $u \in U$ and so $f(u) = g(u)$ for all $u \in U$. ∎

1.4 Difference modules and fibre functors

Let k be a difference field with automorphism ϕ. The ring $k[\Phi, \Phi^{-1}]$ of difference operators consists of the finite sums $\sum_{n \in \mathbf{Z}} a_n \Phi^n$. The multiplication is defined by the formula $\Phi a = \phi(a)\Phi$ with $a \in k$. One considers left modules M

over $k[\Phi, \Phi^{-1}]$ which are of finite dimension over k. Such a module is called a difference module over k. The choice of a basis (e) of M over k identifies M with k^d. On k^d we define ϕ coordinate wise. The induced operator Φ on k^d has the form $\Phi Y = A\phi Y$ where A (the matrix of Φ with respect to the basis (e)) is an invertible matrix with coefficients in k. The difference equation corresponding to M is $\phi Y = A^{-1}Y$. A fundamental matrix for this difference equation is an invertible matrix F satisfying $\phi F = A^{-1}F$. Conversely, given a difference equation $\phi(Y) = A^{-1}Y$ with A an invertible matrix in $Gl(d)(k)$, one can define a difference module structure on k^d via $\Phi Y = A\phi Y$. One sees that two difference equations $\phi(Y) = A_1^{-1}Y$ and $\phi(Y) = A_2^{-1}Y$ define isomorphic modules if and only if there exists $B \in Gl(d)(k)$ such that $A_1^{-1} = \phi(B)A_2^{-1}B^{-1}$. Given a difference module M, one can select a basis and use the associated difference equation to form a Picard-Vessiot extension of k. One sees that a different choice of basis will yield an isomorphic Picard-Vessiot ring. Therefore we can speak of *the* Picard-Vessiot ring for M.

The category of difference modules over k, which will be denoted as $Diff(k, \phi)$, is easily seen to be a rigid abelian tensor category ([20], p.118). More explicitly, for two difference modules M, N one defines its tensor product as $M \otimes_k N$ with a Φ-action defined by $\Phi(m \otimes n) = (\Phi m) \otimes (\Phi n)$. The unit object is the 1-dimensional space ke with $\Phi e = e$. The ring of endomorphisms of the unit object is clearly equal to C the field of constants of k. One could apply Theorem 2.11 of [20] in order to obtain a definition of the difference Galois group of a difference equation over k. To do this we have to construct a covariant fibre functor

$$\omega : Diff(k, \phi) \to Vect_C$$

where $Vect_C$ denotes the category of finite dimensional vector spaces over C. This means that ω is exact, faithful, C-linear, and commutes with tensor products. Theorem 2.11 of [20] then allows one to conclude that ω is an equivalence of tensor categories between $Diff(k, \phi)$ and the category of finite dimensional representations of an affine group scheme G. A natural choice for $\omega(M)$ would be the set of solutions of M in some "solution field" connected with M. As we have seen in Example 0.1 a "solution field" has in general new constants and thus the set of solutions is not a finite dimensional vector space over C. This gives a functorial reason to consider the Picard-Vessiot rings defined above. The results of sections 1.1 and 1.2 allow us to conclude the following.

Theorem 1.32 *Let R be the Picard-Vessiot ring of the difference module M over k.*

1. *Put $V := ker(\Phi - 1, R \otimes_k M)$. Then V is a vector space over C with dimension equal to the dimension of M over k. Let $\{\{M\}\}$ denote the full abelian tensor subcategory of $Diff(k, \phi)$ generated by M and its dual M^*.*

Then the functor

$$\omega_M : \{\{M\}\} \to Vect_C, \ given \ by \ \omega_M(N) := ker(\Phi - 1, R \otimes_k N)$$

is a faithful exact, C-linear tensor functor. In particular $\{\{M\}\}$ is a neutral Tannakian category.

2. *The group G of the automorphisms of R over k commuting with ϕ can be identified with an algebraic subgroup of $Gl(V)$. This group represents the functor $Aut^\otimes(\omega_M)$.*

3. *The fibre functor $\eta_M : \{\{M\}\} \to Vect_k$ given by $\eta_M(N) = N$ as vector space over k induces a representable functor on C-algebras $Hom^\otimes(\omega_M, \eta_M)$. This functor is represented by the G-torsor $Z = spec(R)$.*

4. *Let $Repr(G)$ denote the rigid abelian tensor category of the finite dimensional representations of the linear algebraic group G. The equivalence of the two rigid abelian tensor categories $\{\{M\}\}$ and $Repr(G)$ is explicitly given by the two functors:*

$$\alpha : N \in \{\{M\}\} \mapsto ker(\Phi - 1, R \otimes_k N) \in Repr(G)$$

$$\beta : W \in Repr(G) \mapsto (R \otimes_C W)^G \in \{\{M\}\}$$

5. *$k \subset R$ is the set of G-invariant elements of R.*

Proof: (1) The condition on the existence of a fundamental matrix with coefficients in R implies that V satisfies $R \otimes_k M = R \otimes_C V$. The rest of (1) is straightforward.

(2) The group G acts on $R \otimes_k M$ and this action commutes with the action of Φ on $R \otimes_k M$. Therefore V is invariant under the action of G. Further the coefficients of a basis of V over C expressed in a basis $\{1 \otimes m_i\}$ (where $\{m_i\}$ is a basis of M over k) generate the algebra R over k by the minimality condition on R. This implies that the restriction map $G \to Gl(V)$ is injective. In order to see that G is an algebraic subgroup of $Gl(V)$ we write more explicitly $R = k[X_{i,j}, \frac{1}{D}]/I$ where I is the chosen ideal maximal among all ϕ-invariant ideals. Let $x_{i,j}$ and d denote the images of the $X_{i,j}$ and D in R. Any $g \in G$ has the form $(gx_{i,j}) = (x_{i,j})C(g)$ where $C(g)$ is a matrix with coefficients in C. The map $g \mapsto C(g)$ is in fact the map $G \to Gl(V)$. We have already seen that this is an algebraic subgroup. For any C_k-algebra R the group $G(R)$ coincides with the group of ϕ-equivariant $k \otimes_{C_k} R$-automorphisms of $R \otimes_{C_k} R$. This induces functorial group homomorphisms $\alpha(R) : G(R) \to Aut^\otimes(\omega_M)(R)$. A calculation shows that the $\alpha(R)$ are isomorphisms. This proves (2).

(3) The statement is contained in [20], Theorem (3.2), except for the identification of the G-torsor of $Hom^\otimes(\omega_M, \eta_M)$ with $Z = spec(R)$. This last assertion is easily verified.

(4) The equivalence of the categories $\{\{M\}\}$ and $Repr(G)$ follows from [20], Theorem 2.11, the existence of ω_M and the identification of G with $Aut^\otimes(\omega_M)$. More explicitly, for $N \in \{\{M\}\}$, the C-vector space $ker(\Phi - 1, R \otimes_k N)$ has the same dimension as N over k. The action of G on R commutes with ϕ on R and so G acts on $ker(\Phi, -1, R \otimes_k N)$. Hence $ker(\Phi, -1, R \otimes_k N) \in Repr(G)$.

On the other hand, let W be a representation of G. The space $R \otimes_C W$ is given a Φ-action by the Φ-action on R and the trivial action on W. Further the G-action on $R \otimes W$ is given as the tensor product of the G-actions on R and W. Then $(R \otimes_C W)^G$ is certainly a k-vector space with a Φ-action. In order to see that $(R \otimes_C W)^G$ has the correct dimension over k, one takes a finite extension $L \supset k$ such that $Z(L) \neq \emptyset$. Now $R \otimes_k L = L \otimes_C O(G)$. Then $(R \otimes_C W)^G \otimes_k L =$

$$= ((R \otimes_k L) \otimes_C W)^G = ((O(G) \otimes_C W) \otimes_C L)^G = (O(G) \otimes_C W)^G \otimes_C L$$

We claim that $(O(G) \otimes_C W)^G$ and W have the same dimension over C. To see this note that there is a vector space isomorphism between $(O(G) \otimes_C W)^G$ and $HOM_G(W^*, O(G))$. Therefore it suffices to show that W and $HOM_G(W^*, O(G))$ are isomorphic vector spaces. One does this by showing that the identification of $w \in W$ with the map taking $f \in W^*$ to the regular function $g \mapsto f(wg)$ is an isomorphism.

It follows that the k-vector space $(R \otimes_C W)^G$ has the same dimension as W over C. The fact that α and β preserve the dimensions implies that the natural morphisms $id \to \beta\alpha$ and $id \to \alpha\beta$ are isomorphisms.

(5) Let $\mathcal{E}$ denote the unit object of $\{\{M\}\}$, i.e. $\mathcal{E}$ is a 1-dimensional vector space ke with $\Phi e = e$. Then $\alpha(\mathcal{E})$ is a 1-dimensional vector space with trivial G-action. From (4) we know that $\beta\alpha(\mathcal{E}) = R^G = k$. Note that this gives an alternate proof of Corollary 1.15 ∎

Proposition 1.33 *There exists a k-algebra Ω such that:*

1. *Ω is given an automorphism which extend ϕ on k. This automorphism is also denoted by ϕ.*

2. *Every difference module M over k has a fundamental matrix with coefficients in Ω.*

3. *Ω has only trivial ϕ-invariant ideals.*

4. *Ω is minimal in the sense that no proper subalgebra of Ω satisfies 1,2,3.*

This Ω is unique up to isomorphism. The set of constants of Ω is C. The functor

$$\omega : Diff(k, \phi) \to Vect_C \ \ given \ by \ \omega(M) = ker(\Phi - 1, \Omega \otimes_k M)$$

is a faithful exact, C-linear tensor functor.

Proof: This follows quite easily from Theorem 1.13. Indeed, choose for every difference module M the ring R_M of section 1.1. Choose a family of difference modules $\{M_i\}_{i \in I}$ such that every difference module is a subquotient of some finite sum $M_J := \oplus_{i \in J} M_i$. For any two finite subsets J, J' of I with $J \subset J'$ one can choose a morphism $a(J, J') : R_{M_J} \to R_{M_{J'}}$ such that $a(J, J'') = a(J', J'')a(J, J')$ if $J \subset J' \subset J''$. Then the direct limit Ω of the system $\{R_{M_J}, a(\ , \ )\}$ has the required properties. ∎

The ring Ω of Proposition 1.33 will be called *the universal Picard-Vessiot ring of k*.

Chapter 2

Algorithms for difference equations

2.1 Difference equations of order one

Consider the difference equation $\phi(y) = ay$ with $a \in C(z)^*$, where as usual C is a field of characteristic 0 and ϕ is given by $\phi(z) = z + 1$. One wants to know the difference Galois group of the equation. We will first discuss this from a theoretical point of view and then develop some algorithms.

Let D be a divisor on $\mathbf{P}^1(C) = C \cup \{\infty\}$, i.e. D is a finite formal expression $\sum n_a[a]$ with all $n_a \in \mathbf{Z}$ and the a elements of $\mathbf{P}^1(C)$. The support of a divisor is the finite set of all a with $n_a \neq 0$. As usual, the divisor $div(f)$ of a rational function $f \in C(z)^*$ is given by $div(f) = \sum ord_a(f)[a]$, where the summation is taken over all $a \in \mathbf{P}^1(C)$ and where $ord_a(f)$ denotes the order of f at the point a. We let ϕ act on divisors in the following way $\phi(\sum n_i[a_i]) = \sum n_i[a_i - 1]$. Then clearly $\phi(div(f)) = div(\phi(f))$. It follows that the divisor $D = \sum n_a[a]$ of $\phi(f)f^{-1}$, for any $f \in C(z)^*$, has the properties:

- ∞ is not in the support of D.

- For every $\mathbf{Z}$-orbit E in C, i.e. every subset of C of the form $e + \mathbf{Z}$, one has that $\sum_{a \in E} n_a = 0$.

A divisor of the form $n[a] - n[a - 1]$ with $a \in C$ is the divisor of $\phi(f)f^{-1}$ with $f = (z - a)^{-n}$. Let $D = n_0[\alpha] + n_1[\alpha - m_1] + \ldots n_t[\alpha - m_t]$, $0 < m_1 < m_2 < \ldots < m_t$ be a divisor whose support lies in a $\mathbf{Z}$–orbit and assume that $\sum_{i=1}^{t} n_i = 0$. If we let

$$f = (x - \alpha)^{-n_0}(x - (\alpha - 1))^{-n_0} \cdots (x - (\alpha - (m_1 - 1)))^{-n_0}$$

$$(x - (\alpha - m_1))^{-n_0 - n_1} \cdots (x - (\alpha - (m_2 - 1)))^{-n_0 - n_1}$$
$$(x - (\alpha - (m_2)))^{-n_0 - n_1 - n_2} \cdots (x - (\alpha - (m_3 - 1)))^{-n_0 - n_1 - n_2}$$

$$\vdots \qquad\qquad \vdots \qquad\qquad \vdots$$

$$(x - (\alpha - (m_t - 1)))^{-n_0 - n_1 - \cdots - n_{t-1}} (x - (\alpha - (m_t)))^{-n_0 - n_1 - \cdots - n_t}$$

we have that D is the divisor of $\phi(f)f^{-1}$. Using this, one easily proves that any divisor D having the above two properties is the divisor of $\phi(f)f^{-1}$ for some $f \in C(z)^*$. This rational function f is unique up to a constant in C^*. The following lemma follows at once.

Lemma 2.1 *Let $g \in C(z)^*$. Then g has the form $\phi(f)f^{-1}$ for some $f \in C(z)^*$ if and only if the following three properties hold:*

1. *∞ is not in the support of $div(g)$.*

2. *For every $\mathbf{Z}$-orbit E one has $\sum_{a \in E} ord_a(g) = 0$.*

3. *$g(\infty) = 1$.*

Let $S \subset C$ denote a set of representatives of $C/\mathbf{Z}$, the set of $\mathbf{Z}$-orbits. If C is the field of complex numbers $\mathbf{C}$ then $S = \{s \in \mathbf{C}|\ 0 \leq Re(s) < 1\}$ is such a set of representatives. Using this set of representatives one can define a *standard difference equation* $\phi(y) = ay$ by requiring that $a = c\frac{t}{n}$ where $c \in C^*$ and t, n are relatively prime monic polynomials with g.c.d. one and such that the zeroes of t and n are in S. Two order one difference equations $\phi(y) = a_i y;\ i = 1, 2$ are said to be *equivalent* (that is, they define isomorphic difference modules) if and only if there is a $f \in C(z)^*$ with $a_1 = a_2\phi(f)f^{-1}$.

Lemma 2.2 *Every order one difference equation is equivalent to a unique standard difference equation.*

Proof. Let $\phi(y) = ay$ be a given difference equation. For every $\mathbf{Z}$-orbit $s + \mathbf{Z}$ (with $s \in S$), we define $m_s = \sum_{b \in s + \mathbf{Z}} ord_b(a)$. Let $\frac{t}{n}$, with t and n relatively prime monic polynomials, have divisor $\sum_{s \in S} m_s[s]$. From the above it follows that $a = \phi(f)f^{-1}c\frac{t}{n}$, where $c \in C^*$ and $f \in C(z)^*$. Hence the equation is equivalent to a standard equation. It is easily seen that two distinct standard equations are not equivalent. ∎

Corollary 2.3 *The difference Galois group $G \subset \mathbf{G}_m$ of the standard difference equation $\phi(y) = c\frac{t}{n}y$ is equal to:*

1. *$\{1\}$ if $c = t = n = 1$.*

2. *The finite group μ_k if $t = n = 1$ and c is a primitive k-th root of unity.*

3. $\mathbf{G}_m$ *in all other cases.*

Proof. We know that G is the smallest algebraic subgroup of $\mathbf{G}_m$ such that $\phi(f)f^{-1}c\frac{t}{n} \in G(C(z))$ for some $f \in C(z)$. The corollary follows at once from this. ∎

Standard difference equations are not very convenient for algorithms. In order to bring an equation $\phi(y) = ay$ into standard form one has to know the zeroes and poles of a. We shall now describe an algorithm for calculating the Galois group of $\phi(y) = ay$ that does not require this information. Let us define the *height* of an element $a \in C(z)$ to be $\max(degree(t), degree(n))$ where $a = c\frac{t}{n}$ with $c \in C^*$ and t, n relatively prime monic polynomials.

An order one difference equation $\phi(y) = ay$ is called *minimal* if for every equivalent equation $\phi(y) = by$ one has $height(a) \le height(b)$.

Lemma 2.4 *The following statements are equivalent:*

1. $\phi(y) = c\frac{t}{n}y$ *is minimal.*

2. For every $m \in \mathbf{Z}$, t *and* $\phi^m(n)$ *are relatively prime.*

3. There is no $\mathbf{Z}$*-orbit containing a zero of* t *and a zero of* n.

Furthermore, Corollary 2.3 remains valid if "standard" is replaced by "minimal".

Proof. Clearly, 2. and 3. are equivalent. We now show that $1. \Rightarrow 2$. Assume that 2. does not hold and let $d := g.c.d.(\phi^m(t), n) \ne 1$. We write $t = t_1\phi^{-m}(d)$ and $n = dn_1$. Since $\phi^{-m}(d)d^{-1}$ is of the form $\phi(f)f^{-1}$, this leads to an equivalent equation $\phi(y) = c\frac{t_1}{n_1}y$ of smaller height.

To show that $2. \Rightarrow 1.$, assume that 2. holds. Let $f \in C(x)$. We wish to show that $\frac{\bar{t}}{\bar{n}} = \phi(f)f^{-1}\frac{t}{n}$ has height no higher than the height of $\frac{t}{n}$. Since $g = \frac{\bar{t}n}{\bar{n}t} = \phi(f)f^{-1}$, we have that $\sum_{a \in E} ord_a(g) = 0$ for every $\mathbf{Z}$-orbit E. Since no $\mathbf{Z}$-orbit contains a zero of t and a zero of n, we have that for any $\mathbf{Z}$-orbit E containing a zero of n, $\sum_{a \in E} ord_a(n) = \sum_{a \in E} ord_a(\bar{n})$ and a similar statement concerning t and $\bar{t}$. Therefore $degree(\bar{t}) \ge degree(t)$ and $degree(\bar{n}) \ge degree(n)$, so $height(\frac{\bar{t}}{\bar{n}}) \ge height(\frac{t}{n})$. ∎

Our aim is to transform a general equation $\phi(y) = ay$, with $a = c\frac{t}{n}$, into a minimal equation without using the zeroes and poles of a and without factorization in $C[z]$. The proof of Lemma 2.4 shows us how to proceed. Let C be given as a subfield of the field of complex numbers $\mathbf{C}$ and calculate upper bounds R_1 and

R_2 for the absolute values of the complex zeroes of t and n. Let $N = [R_1 + R_2]$. Let m be an integer with $|m| \leq N$ such that $d := g.c.d.(\phi^m(t), n) \neq 1$. We then write $t = t_1 \phi^{-m}(d)$ and $n = dn_1$. Since $\phi^{-m}(d)d^{-1}$ is of the form $\phi(f)f^{-1}$, this leads to an equivalent equation $\phi(y) = c\frac{t_1}{n_1}y$ of smaller height. If no such m exists, then we know that the equation is minimal since for m with $|m| > N$ the polynomials $\phi^m(t)$ and n can have no common zero. If such an m exists, we repeat the process with the resulting equation of smaller height.

A variation on the method above is the following. Let X denote an indeterminate. The resultant $R(X)$ of the two polynomials $t(z + X)$ and $n(z)$ with respect to the variable z, is a polynomial in X with coefficients in C. An integer m satisfies $R(m) = 0$ if and only if the g.c.d. of $\phi^m(t)$ and n is not one. Again one has only to consider integers m with $|m| \leq R_1 + R_2$.

In the special case $C = \mathbf{Q}$, one can use the factorization in $\mathbf{Q}[z]$ to transform the equation $\phi(y) = ay$ into a minimal one. Let $\{P_1, \ldots, P_R\} \subset \mathbf{Q}[z]$ denote the monic irreducible divisors of the numerator and the denominator of a. Let us call two monic polynomials P and Q equivalent if $Q = \phi^s(P)$ for some integer s. Let $P = z^k + p_{k-1}z^{k-1} + \ldots + p_0$ and $Q = z^k + q_{k-1}z^{k-1} + \ldots + q_0$. If P and Q are equivalent then $Q = \phi^s(P)$ with $sk + p_{k-1} = q_{k-1}$. Hence it is easy to verify whether two polynomials are equivalent. Let the subset $\{P_1, \ldots, P_r\}$ be a set of representatives for the equivalence classes of $\{P_1, \ldots, P_R\}$. Hence the equation is equivalent to an equation $\phi(y) = by$ where $b = cP_1^{n_1} \ldots P_r^{n_r}$ for certain $n_i \in \mathbf{Z}$. This is a minimal equation.

2.2 Difference equations in diagonal form

A diagonal matrix with entries $d_1, \ldots, d_m$ on the diagonal will be denoted by $[d_1, \ldots, d_m]$. One considers a difference equation $\phi(y) = [d_1, \ldots, d_m]y$ where all $d_i \in C(z)^*$. The problems are the same as in the order one case, namely determine the differential Galois group and find a standard form for the equation.

We know that the difference Galois group G is the smallest algebraic subgroup of the m-dimensional torus T such that there exists a $[f_1, \ldots, f_m] \in T(C(z))$ with

$$[\phi(f_1)f_1^{-1}d_1, \ldots, \phi(f_m)f_m^{-1}d_m] \in G(C(z)).$$

A character on T is a homomorphism $\chi : T \to C^*$ of the form $\chi[x_1, \ldots, x_m] = x_1^{n_1} \ldots x_m^{n_m}$ with $\mathbf{n} := (n_1, \ldots, n_m) \in \mathbf{Z}^m$. Let us write $\chi_{\mathbf{n}}$ for the character defined above. Any algebraic subgroup H of T is given as the intersection of the kernels of a number of characters on T. For the determination of the difference Galois group we have to find the $\mathbf{n}$ such that there exists a $[f_1, \ldots, f_m] \in T(C(z))$

with

$$(\phi(f_1)f_1^{-1}d_1)^{n_1}\cdot\ldots\cdot(\phi(f_m)f_m^{-1}d_m)^{n_m} = 1.$$

Suppose that a set of representatives S of the $\mathbf{Z}$-orbits is chosen and that the $d_i = c_i\frac{a_i}{b_i}$ are already in standard form. Then the difference Galois group lies in the kernel of $\chi_{\mathbf{n}}$ if and only if

$$c_1^{n_1}\cdot\ldots\cdot c_m^{n_m} = 1 \text{ and } (\frac{a_1}{b_1})^{n_1}\cdot\ldots\cdot(\frac{a_m}{b_m})^{n_m} = 1.$$

The last equation can be translated into $n_1 div(d_1) + \ldots + n_m div(d_m) = 0$. Assuming as before that the divisors of the d_i are known then one can with linear algebra determine the submodule $V \subset \mathbf{Z}^m$ consisting of all $\mathbf{n}$ satisfying $n_1 div(d_1) + \ldots + n_m div(d_m) = 0$. It seems more difficult to give a general procedure to solve the first equation.

For $C = \mathbf{Q}$ we will give an algorithm which determines the difference Galois group of the equation $\phi(y) = [d_1, \ldots, d_m]y$, using factorization in $\mathbf{Q}[z]$ and $\mathbf{Z}$. Let $\{P_1, \ldots, P_R\}$ denote the monic irreducible polynomials in $\mathbf{Q}[z]$ which occur in the numerators and the denominators of the d_i. As above two monic polynomials P and Q are called equivalent if $Q = \phi^s(P)$ for some integer s. Let the subset $\{P_1, \ldots, P_r\}$ be a set of representatives for the equivalence classes of $\{P_1, \ldots, P_R\}$. Then one can transform the original diagonal equation such that every d_i has the form $c_i P_1^{A(i,1)}\cdot\ldots\cdot P_r^{A(i,r)}$. We note that for any $\mathbf{n}$ the element $d_1^{n_1}\cdot\ldots\cdot d_m^{n_m}$ is already in minimal form. So G lies in the kernel of $\chi_{\mathbf{n}}$ if and only if $d_1^{n_1}\cdot\ldots\cdot d_m^{n_m} = 1$.

Let $p_1, \ldots, p_s$ denote the primes occurring in the numerators and the denominators of the c_i. Take suitable $C(i) \in \{0,1\}$ then

$$d_i = (-1)^{C(i)}p_1^{B(i,1)}\cdot\ldots\cdot p_s^{B(i,s)} P_1^{A(i,1)}\cdot\ldots\cdot P_r^{A(i,r)}.$$

The $\mathbf{n}$ such that G lies in the kernel of $\chi_{\mathbf{n}}$ form a subgroup V of $\mathbf{Z}^m$. Further $\mathbf{n} \in V$ if and only if $\mathbf{n}$ satisfies the equations:

(i) $\sum_i n_i A(i,j) = 0$ for $j = 1, \ldots, r$.

(ii) $\sum_i n_i B(i,j) = 0$ for $j = 1, \ldots, s$.

(iii) $\sum_i n_i C(i) \equiv 0 \bmod 2$.

The subgroup V is easily computed and so G is known. Let W denote the subgroup of the $\mathbf{n} \in \mathbf{Z}^m$ such that a nonzero multiple of $\mathbf{n}$ lies in V. Then clearly $[W : V]$ is 1 or 2. This means that G/G^o is either trivial or has two elements.

2.3 Difference equations of order two

We will sketch here an algorithm. A full description can be found in [26]. A difference module of dimension two over $C(z)$ has a cyclic vector as one easily sees. Therefore we may work with order two equations in the form

$$(\phi^2 + a\phi + b)y = 0 \text{ with } a, b \in C(z) \text{ and } b \neq 0.$$

The corresponding matrix equation is

$$\phi(y) = Ay \text{ with } A = \begin{pmatrix} 0 & 1 \\ -b & -a \end{pmatrix}.$$

The difference Galois group of the equation is denoted by G.

Step 1: determine whether the group G is reducible.

If G is reducible the the operator $y \mapsto A^{-1}\phi(y)$ on the space $C(z)^2$ must fix a line. The line $C(z) \begin{pmatrix} 0 \\ 1 \end{pmatrix}$ is not fixed. So we may suppose that the fixed line is $C(z) \begin{pmatrix} 1 \\ u \end{pmatrix}$. From $A^{-1}\phi \begin{pmatrix} 1 \\ u \end{pmatrix} = f \begin{pmatrix} 1 \\ u \end{pmatrix}$ one concludes that

$$u\phi(u) + au = -b.$$

This equation will be called *the Riccati equation*. We write this equation as $A_1 u\phi(u) + A_2 u = A_3$ where $A_1, A_2, A_3 \in C[z]$ satisfy g.c.d.$(A_1, A_2, A_3) = 1$. Write $u = \frac{T}{N}$ with $T, N \in C[z]$, N a monic polynomial and g.c.d.$(T, N) = 1$. Let P denote the greatest monic divisor of N such that $\phi(P)$ divides T, in other words $P = g.c.d.(\phi^{-1}(T), N)$. Then one can write $u = c\frac{\phi(P)}{P}\frac{t}{n}$ where $c \in C^*$, P, t, n are monic polynomials, g.c.d.$(t, \phi(n)) = 1$ and g.c.d.$(\phi(P)t, Pn) = 1$. The equation now reads as

$$A_1\phi^2(P)t\phi(t)c^2 + A_2\phi(P)\phi(n)t = A_3 n\phi(n)P.$$

It follows that t is a monic divisor of A_3 and that n is a monic divisor of $\phi^{-1}(A_1)$. This gives finitely many possibilities for $\frac{t}{n}$.

One continues by calculating the possible solutions of the Riccati equation in $C((z^{-1}))$. In fact one only needs a truncation of such a possible solution. There are at most two truncated solutions. From the truncated solution one can determine the degree d of the monic polynomial P and the constant c. Write $P = z^d + x_{d-1}z^{d-1} + \ldots + x_0$ with indeterminates $x_{d-1}, \ldots, x_0$. Combining the possibilities for $\frac{t}{n}$ and the d one finds a set of linear equations for the x_i. In this way all solutions of the Riccati equation are found. This ends the algorithm.

We remark that this method may give rise to a quadratic field extension of the base field C, which is a priori not algebraically closed. The possibilities for this quadratic extension can be found in advance, i.e. without trying to solve the Riccati equation.

If the algorithm above gives a solution of the Riccati equation, then one can proceed with a difference equation $\phi(y) = Ay$ where the matrix A is an upper triangular matrix. If we have found two independent solutions of the Riccati equation then A is moreover a diagonal matrix. Using the earlier methods for order one and equations in diagonal form, one can determine the difference Galois group. If we have found only one solution of the Riccati equation then one can treat the diagonal part of the matrix A as before and determine the difference Galois group.

If the algorithm does not produce solutions of the Riccati equation then the group G is irreducible. The next step is to see whether G is an imprimitive group.

Step 2: determine if G is an imprimitive group.

If G happens to be imprimitive then there is a $B \in Gl(2, C(z))$ such that $B^{-1}A\phi(B) \in G(C(z))$. Then $B^{-1}A\phi(B)$ permutes two lines. Using this one finds a Riccati equation like the one studied above but with ϕ replaced by ϕ^2. If the algorithm gives a solution then one can transform the equation such that the new equation $\phi(y) = Ay$ has a matrix of the form $\begin{pmatrix} 0 & * \\ * & 0 \end{pmatrix}$. A classification of the imprimitive groups H such that H/H^0 is a cyclic group leads to a complete determination of the difference Galois group, to a standard difference equation and to symbolic solutions.

If the algorithm shows that G is not imprimitive then we proceed as follows.

The final step.

We know that G is irreducible and not imprimitive and that G/G^0 is a cyclic group. This implies that $Sl(2)_C \subset G \subset Gl(2)_C$. One determines G by solving the order one difference equation $\phi(y) = det(A)y$. This ends the algorithm.

For further algorithms concerning difference equations, we refer the reader to [1, 2, 43, 44, 45, 46].

Chapter 3

The inverse problem for difference equations

The field C is supposed to be algebraically closed of characteristic 0 and $C(z)$ is made into a difference field by $\phi(z) = z + 1$. The theme of this section is the conjecture:

Conjecture *An algebraic subgroup G of $Gl(d)_C$ is the difference Galois group of a difference equation $\phi(Y) = AY$ over $k = C(z)$ if and only if G/G° is cyclic.*

Remark: As we have seen in Proposition 1.20 the condition G/G° cyclic is necessary.

We note that for differential equations over $\mathbf{C}(z)$ any linear algebraic subgroup of $Gl(d)$ is the Galois group of a differential equation of order d over $\mathbf{C}(z)$. The well known proof, based on the Riemann-Hilbert correspondence, is analytic in nature (c.f., [53], [60]). However for a connected linear algebraic group G over any algebraically closed field C of characteristic 0, there is a recent constructive and purely algebraic proof that G is a differential Galois group (see [41]).

In Chapter 8 we will discuss the relationship between the difference Galois group and the connection matrix for a difference equation over the fields $\mathbf{C}(z)$ and $\mathbf{C}(\{z^{-1}\})$. This leads to an analytic proof of this conjecture when G is connected or a cyclic extension of a torus. In this section we will give an algebraic (and constructive) proof of the following result.

Theorem 3.1 *Any connected algebraic subgroup G of $Gl(d)_C$ is the difference Galois group of a difference equation $\phi(Y) = AY$ over $k = C(z)$.*

Proposition 3.2 *Consider the difference equation $\phi(Y) = AY$ with $A \in Gl(d)_k$, where $k := C(z)$ and let $G \subset Gl(d)_C$ be the Galois group of $\phi(Y) = AY$. Let T or T_A denote the smallest algebraic subgroup of $Gl(d)_C$ such that $A \in T(k)$. Then:*

1. *Let U denote the open subset of $\mathbf{P}^1(C)$ consisting of the elements a with $A(a) \in Gl(d)(C)$. Then T is generated as an algebraic subgroup by $\{A(a)|a \in U\}$.*

2. *G is (after conjugation) a subgroup of T and $\dim T \leq 1 + \dim G$.*

Proof. 1. Let S denote the algebraic subgroup of $Gl(d)_C$ generated by all $A(a)$ with $a \in U$. Clearly $A(a) \in T(C)$ for $a \in U$. Hence $S \subset T$. The element A can be seen as a rational map $\mathbf{P}^1_C \to S$ defined on U. Thus $A \in S(k)$ and $S = T$.

2. On $Gl(d)_k$ we consider the map τ given by $\tau(B) = A^{-1}\phi(B)$. We recall that we are trying to find a minimal τ-invariant Zariski-closed subset Z of $Gl(d)_k$. For any such minimal Z one has by Proposition 1.20 that $Z = B_0 G_k$ for some $B_0 \in Gl(d)(k)$ and with G the difference Galois group of the equation. The set T_k is clearly τ-invariant. So we may suppose that $Z \subset T_k$ and one has clearly $G \subset T$.

One considers the (quasi-projective) variety T/G. Then τ operates on $(T/G)_k$ in the natural way, i.e. $\tau(BG_k) = A^{-1}\phi(B)G_k$. The point $B_0 G_k$ is a fixed point for this action of τ. Let $Y \subset T/G$ denote the smallest Zariski-closed subset such that $B_0 G_k \in Y(k)$. Then Y is either a point or a rational curve on T/G. Indeed, as in 1., above, the element $B_0 G_k$ is seen as a rational map $\mathbf{P}^1_C \to T/G$, given by $p \mapsto B_0(p)G$. Therefore Y is the Zariski-closure of the image of this map. Let W denote the preimage of Y in T. It follows that $Z \subset W_k$, W_k is τ-invariant and $\dim W_k \leq 1 + \dim Z$. After multiplying both W and Z on the right handside by a suitable $B \in T(C)$ we may suppose that $1 \in W$. The τ-invariance of W_k implies that for every $B \in W(C)$ and every $a \in U$ one has $A(a)B \in W(C)$. Hence W is a right coset for T. But since $W \subset T$ one has $W = T$. ∎

We note that Proposition 3.2 was motivated by a similar statement for *differential* Galois groups given by Magid as Theorem 7.13 in [39]. Regrettably this latter Theorem is incorrect (see [41]).

Proposition 3.3

1. *For every $A \in Gl(d)(k)$ the group $T = T_A$ satisfies T/T^o is cyclic.*

2. *Let G be an algebraic subgroup of $Gl(d)_C$ such that G/G^o is cyclic. Then there is a $A \in G(k)$ such that $T_A = G$.*

3. The difference Galois group of the $\phi(Y) = AY$ as in 2. is a subgroup of G of codimension less than or equal to 1.

Proof. 1. Let BT° denote the coset of T/T° such that $A \in BT^\circ(k)$. Let T' denote the subgroup of T generated by T° and $B \in T(C)$. Then $A \in T'(k)$. Hence $T = T'$ and T/T° is cyclic.

2. Let us assume we have found an element $A \in G^0(k)$ such that $T_A = G^0$, that is a rational map $A : \mathbf{P}_C^1 \to G^0$. One can suppose that A is defined at $z = 0$ and that $A(0) = id$. Let $B \in G(C)$ be an element whose coset generates G/G^0. If we let $N = BA$, then $N(0) = B \in T_N$ and so $T_A \subset T_N$. This implies that $T_N = G$. This observation reduces the proof of 2. to the case where G is connected. One knows that G is a rational variety. Let $U \subset G$ be a Zariski-open subset isomorphic to an open subset of affine space over C. One can find a finite number of elements $a_1, ..., a_s \in U$ such that G is generated as an algebraic group by $a_1, ..., a_s$ (c.f., [60]). There is a rational curve in U passing through the points $a_1, ..., a_s$. This rational curve defines as an element $A \in U(k) \subset G(k)$. Now A has the required property.

3. is a consequence of Proposition 3.2 ∎

Remarks. 1) In [26] (see Chapter 2.3) one has developed a "Kovacic algorithm" for difference equations over $C(z)$ of order ≤ 2. Using this algorithm one can verify that all algebraic subgroups G of $Gl(2)_C$ with cyclic G/G° occur as difference Galois groups.

2) $Gl(n, C)$ with $n \geq 3$ is a difference Galois group. Indeed, take a number of matrices $A_0, ..., A_s$ generating $Gl(n, C)$ as algebraic group. By elementary interpolation there exists a polynomial map $A : C \to M(n, C)$ with $A(i) = A_i$ for $i = 0, ..., s$. Then $A \in Gl(n, k)$ and the difference Galois group of A is a subgroup of $Gl(n, C)$ of codimension ≤ 1. If this group is strictly smaller than $Gl(n, C)$ then it must be a finite extension of $Sl(n, C)$. One can avoid those finite extensions by multiplying A by a general polynomial $f \in C[z]$ such that $f(i) = 1$ for $i = 0, ..., s$.

3) A simple linear algebraic group G different from $Sl(2, C)$ and $PSl(2, C)$ has no algebraic subgroups of codimension 1. The Proposition 3.2 implies that such a group is a difference Galois group.

4) In the next lemma we prove that any $\mathbf{G}_a^m(C)$ (where $\mathbf{G}_a$ is the additive group) and any torus over C is a difference Galois group. We have included a somewhat stronger statement which will be useful in the proof of the theorem.

Lemma 3.4

1. Let V be a finite dimensional vector space over C. Let the action of ϕ on $k \otimes_C V$ be given by $\phi(a \otimes v) = (\phi a) \otimes v$. Let L be a k-linear automorphism of $k \otimes V$. There exists $m \in k \otimes_C V$ such that for any proper subspace $W \subset V$, such that $k \otimes W$ is invariant under L, and any $b \in k \otimes_C V$ one has $-m + L(\phi(b)) - b \notin k \otimes_C W$.

2. Any $\mathbf{G}_a^n(C)$ is a difference Galois group.

3. Any torus G over C is a difference Galois group.

Proof. 1. Any element of k has a partial fraction decomposition of the form

$$P(z) + \sum_{\alpha \in C; n \geq 1} \frac{c(n, \alpha)}{(z - \alpha)^n},$$

where $P(z)$ is a polynomial and where $c(n, \alpha) \in C$. An element $a \in k \otimes V$ can be given a similar decomposition

$$P(z) + \sum_{\alpha \in C; n \geq 1} \frac{v(n, \alpha)}{(z - \alpha)^n},$$

where $P(z) \in C[z] \otimes V$ and $v(n, \alpha) \in V$. This decomposition is unique. We will say that a has a pole at α if some $v(n, \alpha) \neq 0$. The order of the pole of a at α is the maximal n such that $v(n, \alpha) \neq 0$.

For a suitable monic polynomial $f \in C[z]$ the map L can also be seen as an automorphism of $R \otimes_C V$ with $R = C[z, \frac{1}{f}]$. Take a finite set of elements $v_1, \ldots, v_r \in V$ such that for every proper subspace W of V with $L(k \otimes W) = k \otimes W$, one has $\{v_1, \ldots, v_r\} \not\subset W$. One could take for $\{v_1, \ldots, v_r\}$ a basis of V. If L is the identity then this is a minimal choice. For other automorphisms L a smaller set is sometimes possible.

Let $\beta_1, \ldots, \beta_s$ denote the zeroes of f. Choose elements $\alpha_1, \ldots, \alpha_r \in C$ such that the images of the α_i in $C/\mathbf{Z}$ are distinct and different from the images of the β_j in $C/\mathbf{Z}$. For m we choose $\sum_i \frac{1}{z - \alpha_i} v_i$. Suppose that this element does not satisfy the property required in 1. Then there exists a $b = b(z) \in k \otimes V$ and a proper subspace $W \subset V$ such that $k \otimes W$ is L invariant and $-m - b(z) + L(b(z + 1)) \in k \otimes W$. We divide by the subspace W and find the equation

$$\bar{L}(\bar{b}(z + 1)) - \bar{b}(z) = \bar{m} \text{ in the space } k \otimes V/W.$$

In this expressions $\bar{L}, \bar{b}$ and $\bar{m}$ stand for the images in $k \otimes V/W$. We note that $\bar{L}$ can also be seen as an automorphism of $R \otimes V/W$. Further $\bar{m}$ can have at most poles of order 1 for the α_i and it has no other poles. Since some $v_i \notin W$ we find that $\bar{m} \neq 0$ and has for some α_i a pole of order 1.

We want to investigate the poles of $\bar{b}$ at the points $\alpha_i + \mathbf{Z}$. By construction, $\bar{L}(\bar{b})$ has a pole at $\alpha_i + n$ (with $n \in \mathbf{Z}$) if and only if $\bar{b}$ has a pole in $\alpha_i + n$. From the equation it follows that $\bar{b}$ has a pole in α_i or $\alpha_i + 1$. In the first case all $\alpha_i - 1, \alpha_i - 2, ..$ are poles of $\bar{b}$. In the second case all $\alpha_i + 1, \alpha_i + 2, \alpha_i + 3, ...$ are poles of $\bar{b}$. This is a contradiction since b and $\bar{b}$ have only finitely many poles.

2. One can identify $\mathbf{G}_a(C)^n$ with V and the algebraic subgroups of $\mathbf{G}_a(C)^n$ with vector spaces $W \subset V$. Now 2., with the identity for L, proves 1.

3. Let $\chi_1, ..., \chi_n$ denote a basis of the characters of G. Choose $A \in G(k)$ such that the elements $\chi_i(A) \in K^*$ are multiplicatively independent modulo the subgroup $\{\frac{\phi f}{f} \mid f \in C(z)^*\}$. For example, if we identify $G(C)$ with $(C^*)^t$, and let χ_i be the ith coordinate function, then we can let $A = (x - 1, x - \frac{1}{2}, ..., x - \frac{1}{t})$. The elements $\chi_i(A)$ are multiplicatively independent modulo the above subgroup because an element of the form $\prod_{i=1}^{t}(x - \frac{1}{i})^{n_i} = \frac{\phi(f)}{f}$ only if each $n_i = 0$ (c.f., Proposition 2.1) For such a selection of A, only the trivial character of G can have the value 1 on $B^{-1}A^{-1}\phi(B)$, where B is any element of $G(k)$. For every proper subgroup H of G there is a character $\chi \neq 1$ with $\chi(H) = 1$. Proposition 1.21 shows that the Galois group must be all of G. ∎

The proof of Theorem 3.1 Consider the following statements:

(1) Let the linear algebraic group G over C have the form

$$(G_1)^{n_1} \times (G_2)^{n_2} \times ... \times (G_r)^{n_r} \times T$$

where the G_i are distinct, simply connected, noncommutative simple group and T is a torus. Then G is a difference Galois group.

(2) Every reductive connected linear algebraic group is a difference Galois group.

(3) Every connected linear algebraic group with commutative unipotent radical is a difference Galois group.

(4) Every connected linear algebraic group is a difference Galois group.

We will show $(1) \Rightarrow (2) \Rightarrow (3) \Rightarrow (4)$ and then give a proof of (1).

Lemma 3.5 $(1) \Rightarrow (2)$.

Proof. For a connected reductive group G there exists a group G' satisfying (1) and a surjective morphism $\pi : G' \to G$ of algebraic groups with a finite kernel N. Let $A' \in G'(k)$ be chosen such that G' is the difference Galois group of the equation $\phi(Y) = A'Y$. This A' induces an action τ on $G'(\bar{k})$ given by $\tau(B) = (A')^{-1}\phi(B)$ and a corresponding automorphism ϕ of $O(G'_k)$. For $Gl(d)$ this is explained in Section 1.2. For $G' \subset Gl(d)$, τ on $G'(\bar{k})$ is the restriction of the τ on $Gl(d)(\bar{k})$ and ϕ on $O(G'_k) = O(Gl(d)_k)/I$, where I is the ideal of G', is induced by ϕ on $O(Gl(d)_k)$. The relation between τ and ϕ is again given by the formula

$$(\phi f)(\tau B) = \phi(f(B)) \text{ for all } f \in O(G'_k) \text{ and } B \in G'(\bar{k}).$$

The statement that G' is the difference Galois group of $\phi(Y) = A'Y$ is equivalent to $O(G'_k)$ has no proper ϕ-invariant ideals.

The map π induces a finite injective map $O(G_k) \to O(G'_k)$. In fact $O(G_k)$ is equal to the algebra of the $f \in O(G'_k)$ satisfying $f(Bn) = f(B)$ for all $n \in N$. The subalgebra $O(G_k)$ is invariant under the action of ϕ on $O(G'_k)$. Indeed, for $n \in N$ and $B \in G'(\bar{k})$, one has

$$(\phi f)((\tau B)n) = (\phi f)(\tau(Bn)) = \phi(f(Bn)) = \phi(f(B)) = (\phi f)(\tau B),$$

and τ is bijective on $G'(\bar{k})$. The map τ on $G(\bar{k})$ corresponding to the restriction of ϕ to $O(G_k)$ is easily seen to be $\tau(B) = A^{-1}\phi(B)$, where $A := \pi(A')$.

Suppose now that G is not the difference Galois group of the equation $\phi(Y) = AY$. Then $O(G_k)$ has a non trivial ϕ-invariant ideal J. Since $O(G_k) \to O(G'_k)$ is finite, the ideal $JO(G'_k)$ is a non-trivial ϕ-invariant ideal. This contradicts the hypothesis.   ∎

Lemma 3.6 $(2) \Rightarrow (3)$.

Proof. Consider the morphism $\pi : G \to G/R_u$. By assumption, the unipotent radical R_u is commutative and can therefore be identified as a vector space V over C. On V we will use additive notation and sometimes multiplicative notation for the group law.

The reductive group G/R_u acts on V by conjugation. One knows that there is a subgroup $P \subset G$ which is mapped isomorphically to G/R_u. Such a subgroup is called a Levi-factor. The group G is a semi-direct product of V and P. Any other Levi-factor Q has the form vPv^{-1} for a certain $v \in V$. We investigate first the (algebraic) subgroups H of G which are mapped under π onto G/R_u. The kernel $H \cap V$ of the restriction of π to H is clearly the unipotent radical of H. A Levi-factor $Q \subset H$ for H is also a Levi-factor for G. Hence there exists a $v \in V$ with $vHv^{-1} \supset P$.

We identify the group G with the semidirect product of P and V. In this way, we can write $O_k(G) = k[p, v]$ for some $p \in P, v \in V$. We choose now an $A = A_0 A_1 \in G(k)$ with $A_0 \in P(k)$ and $A_1 \in V(k)$. The A_0 is chosen such that P (or G/R_u) is the difference Galois group of A_0. The choice A_1 will be specified later. We define a difference structure on $O_k(G)$ by setting $\phi(p) = A_1 p$ and $\phi(pv) = Apv$. We will select A_2 in such a way that $O_k(G)$ has no ϕ-invariant ideals or, equivalently, that the Galois group of $\phi(Y) = AY$ is G. If J is a maximal ϕ-invariant ideal of $O_k(G)$, then, by assumption $J \cap k[p] = (0)$. Therefore, we may identify $k[p]$ with a subring of the Picard-Vessiot extension of $\phi(Y) = AY$. This implies that the difference Galois group H of the equation $\phi(Y) = AY$ maps surjectively to G/R_u and, by the previous paragraph, $v^{-1}Hv \supset P$ for a certain $v \in V$. Moreover there is a $B \in G(k)$ with $B^{-1}A^{-1}\phi(B) \in H(k)$. After replacing B by Bv we may suppose that $H \supset P$. The group H is a semi-direct product of P with some subspace $W \subset V$ which is invariant under the action of G/R_u on V by conjugation. Write $B = B_1 B_0$ with $B_0 \in P(k)$ and $B_1 \in V(k)$. Since B_0 and $\phi(B_0)$ are in $P(k)$ we have also $B_1^{-1}A^{-1}\phi(B_1) \in H(k)$. Then $B_1^{-1}A_1^{-1}A_0^{-1}\phi(B_1)A_0 \in W(k)$. Now we will use the additive notation for the group V. Write b for $B_1 \in V(k)$ and m for $A_1 \in V(k)$ and L for the k-linear operator on $k \otimes V$ given by the conjugation with A_0^{-1}. The space $k \otimes W$ is invariant under L and b is such that

$$L(\phi(b)) - b - m \in k \otimes W.$$

From the first part of Lemma 3.4 it follows that there is a choice of m (that is, a choice of $A_1 \in V(k)$) such that the equation has no solution b if W is a proper subspace of V. Hence for this choice of A_1 the difference Galois group corresponding to $A = A_0 A_1$ is equal to G. This finishes the proof. ∎

Lemma 3.7 $(3) \Rightarrow (4)$.

Proof. Let (R_u, R_u) denote the commutator subgroup of the unipotent radical R_u of the connected linear algebraic group G. The canonical map $\pi : G \to G' := G/(R_u, R_u)$ has the property that any element $A' \in G'(k)$ lifts to an element $A \in G(k)$. For A' we choose an element with difference Galois group G'. The difference Galois group $H \subset G$ of the equation $\phi(Y) = AY$ maps surjectively under π to G'. Kovacic ([34], Lemma 2) has shown that this implies $H = G$. ∎

Lemma 3.8   1. *If G_1, G_2 are linear algebraic groups over C such that the only common homomorphic image of both groups is $\{1\}$, then the only algebraic subgroup $G \subset G_1 \times G_2$ which maps surjectively to both factors is $G_1 \times G_2$.*

2. If G_1 and G_2 are difference Galois groups such that the only common homomorphic image of both groups is $\{1\}$, then $G_1 \times G_2$ is a difference Galois group.

Proof. 1. Let $\pi_i : G \to G_i$ denote the two projections. The kernel of π_1 has the form $\{1\} \times K$ where K is a closed subgroup of G_2. Since π_2 is surjective, the group K is normal in G_2. The map $G \overset{\pi_2}{\to} G_2 \to G_2/K$ factors over the kernel of π_1. Thus we find a surjective morphism $G_1 \to G_2/K$. By assumption $G_2/K = \{1\}$. Hence $G = G_1 \times G_2$.

2. We choose $A_i \in G_i(k)$ such that G_i is the difference Galois group of the equation $\phi(Y) = A_i Y$. Let $G \subset G_1 \times G_2$ be the difference Galois group of the equation $\phi(Y) = AY$ with $A = A_1 \times A_2$, then G maps surjectively to the G_i. Hence 2. follows from 1. ∎

Lemma 3.9 For any $m \geq 1$ the group $Sl(2)_C^m$ is a difference Galois group.

Proof. We start by considering the most complicated case $m = 1$. Let $a \in C[z]$ be a nonconstant polynomial satisfying $a(0) = 0$. We consider the difference Galois group $G \subset Sl(2)_C$ corresponding to the difference equation $\phi(Y) = AY$ with

$$A = A(z) = \begin{pmatrix} 0 & -1 \\ 1 & a \end{pmatrix}.$$

We will first show that the smallest algebraic subgroup T of $Sl(2)_C$ such that $A \in T(k)$ is equal to $Sl(2)_C$. The group T is generated as an algebraic group by $\{A(c) | c \in C\}$. One sees that $A(0)^3 A(z) = \begin{pmatrix} 1 & a \\ 0 & 1 \end{pmatrix}$. Hence T contains the matrices $\begin{pmatrix} 1 & c \\ 0 & 1 \end{pmatrix}$ with $c \in C$. Conjugation with $A(0)$ proves that T contains also the matrices $\begin{pmatrix} 1 & 0 \\ c & 1 \end{pmatrix}$ for all $c \in C$. Hence $T = Sl(2)$. From Proposition 3.3 we conclude that G has codimension ≤ 1 in $Sl(2)_C$. Suppose now that $G \neq Sl(2)_C$. Then G is a Borel subgroup of $Sl(2)_C$. By Proposition 1.21 there exists a $B \in Sl(2, k)$ such that $B^{-1}A^{-1}\phi(B) \in G(k)$. After conjugation with a matrix in $Sl(2, C)$ we may suppose that $B^{-1}A^{-1}\phi(B) \in G(k)$ has the form $\begin{pmatrix} * & * \\ 0 & * \end{pmatrix}$. Let e_1, e_2 denote the canonical base of k^2, then e_1 is an eigenvector of $B^{-1}A^{-1}\phi(B)$. Put $v = Be_1$ then $\phi(v) = fAv$ for some $f \in k$. Clearly $v \notin ke_1$, and $v \notin ke_2$. We may therefore write $v = e_1 + ue_2$ for some $u \in k$. The linear dependence of $\phi(v)$ and Av leads to an equation $u\phi(u) + au + 1 = 0$ satisfied by u. This equation can also be written as

$$\phi(u) + u^{-1} = -a.$$

This equation is the analogue for difference equations of the Riccati equation for differential equations. We will show that the equation above has no solution in k. This will finish the proof of the case $m = 1$.

Write $u = \frac{T}{N}$ with $T, N \in C[z]$ such that g.c.d.$(T, N) = 1$ and N is a monic polynomial. Let P denote the greatest monic divisor of N such that $\phi(P)$ divides T. Then one has $u = c\frac{\phi(P)}{P}\frac{t}{n}$ where $c \in C^*$, P, t, n are monic polynomials, g.c.d.$(t, \phi(n)) = 1$ and g.c.d.$(\phi(P)t, Pn) = 1$. The equation reads now:

$$\phi^2(P)\phi(t)tc + Pn\phi(n)c^{-1} = -a\phi(P)\phi(n)t.$$

It follows that $t = 1$ and $n = 1$. The equation simplifies into $c\phi^2(P) + c^{-1}P = -a\phi(P)$. For any monic polynomial P the right-hand side has a degree which is strictly greater than the degree of the left-hand side. This shows that there is no solution $u \in k^*$ of the equation.

The case $m > 1$ will be done using induction on m. We choose for the element $A \in Sl(2, k)^m$ the following

$$A = A(z) = \begin{pmatrix} 0 & -1 \\ 1 & z \end{pmatrix} \times \begin{pmatrix} 0 & -1 \\ 1 & z^2 \end{pmatrix} \times \dots \times \begin{pmatrix} 0 & -1 \\ 1 & z^m \end{pmatrix}.$$

Let $G \subset Sl(2)_C^m$ denote the difference Galois group of the equation $\phi(Y) = AY$. We want to show that the smallest algebraic subgroup $T \subset Sl(2)_C^m$ such that $A \in T(k)$ is $T = Sl(2)_C^m$.
One sees that

$$A(0)^3 A(z) = \begin{pmatrix} 1 & z \\ 0 & 1 \end{pmatrix} \times \begin{pmatrix} 1 & z^2 \\ 0 & 1 \end{pmatrix} \times \dots \times \begin{pmatrix} 1 & z^m \\ 0 & 1 \end{pmatrix}.$$

The smallest algebraic subgroup of $Sl(2)_C^m$ containing all $A(0)^m A(c)$ is of course $\mathbf{G}_a^m$ where $\mathbf{G}_a$ is identified with the subgroup $\begin{pmatrix} 1 & * \\ 0 & 1 \end{pmatrix}$ of $Sl(2)_C$. Thus $\mathbf{G}_a^m \subset T$. By conjugation with $A(0)$ one finds that also

$$\begin{pmatrix} 1 & 0 \\ * & 1 \end{pmatrix} \times \begin{pmatrix} 1 & 0 \\ * & 1 \end{pmatrix} \times \dots \times \begin{pmatrix} 1 & 0 \\ * & 1 \end{pmatrix}$$

lies in T. Hence $T = Sl(2)_C^m$. From Proposition 3.3 one concludes that G has codimension ≤ 1 in $Sl(2)_C^m$. Let $\pi : Sl(2)_C^m \to Sl(2)_C^{m-1}$ denote the projection on the first $m - 1$ factors. By induction $\pi(G) = Sl(2)_C^{m-1}$. The kernel of the restriction of π to G has the form $1 \times \dots \times 1 \times K$, where $K \subset Sl(2)_C$ is a subgroup of codimension ≤ 1. The projection of G to the last factor of $Sl(2)_C^m$ is surjective according to the case $m = 1$. Therefore K is a normal subgroup of $Sl(2)_C$ of codimension ≤ 1. This proves $G = Sl(2)_C^m$ and ends the proof of the proposition. ∎

The proof of (1). Let G be a group described in (1). Lemmas 3.8, 3.4, and 3.9 reduce the general case to $G = H^m$ where H is a simple and simply connected, noncommutative algebraic group over C, distinct from $Sl(2)_C$. For such a group, one can show that H has no subgroups of codimension 1. By induction on m one can see that a proper subgroup G of H^m has codimension ≥ 2. Indeed, consider the map $\pi : H^{m+1} \to H^m$ which forgets the last factor. If $G \subset H^{m+1}$ has codimension ≤ 1, then $\pi(G)$ has codimension ≤ 1 and so the restriction π' of π to G is surjective. The kernel of π' has the form $1 \times 1 \times \ldots \times K$ with K a closed subgroup of H. This closed subgroup has again codimension ≤ 1. hence $K = H$ and $G = H^{m+1}$. An application of Proposition 3.3 ends the proof. ∎

Chapter 4

The ring $\mathcal{S}$ of sequences

The ring $\mathcal{S}$ was defined in Example 1.3 as being the difference ring of germs at infinity of C−valued functions on the nonnegative integers, where C is an algebraically closed field and ϕ_0 is the shift automorphism. We shall show that if k is a perfect difference subfield of $\mathcal{S}$ whose algebraic closure $\bar{k}$ also lies in $\mathcal{S}$ and $\phi_0(Y) = AY$ is a difference equation over k with $det(A) \neq 0$, then the associated Picard-Vessiot ring can be embedded in $\mathcal{S}$. We will then use this fact to prove several conjectures concerning sequences that satisfy linear difference equations over (C, ϕ_0) and, when C has characteristic zero, $(C(z), \phi_0)$. We note that the condition that $\bar{k}$ lies in $\mathcal{S}$ is satisfied by any field $k \subset \mathbf{C}(\{z^{-1}\})$. To see this note that we can embed the algebraic closure of $\mathbf{C}(\{z^{-1}\})$ into $\mathcal{S}$ by selecting branch for each $t^{\frac{1}{m}} = (z^{-1})^{\frac{1}{m}}$ that is real and positive on the positive real axis and mapping each $t^{\frac{1}{m}}$ to the sequence defined by evaluating this branch at sufficiently large integers. In particular, $(C(z), \phi_0)$ satisfies our hypothesis. In the sequel we will sometimes write ϕ instead of ϕ_0.

Proposition 4.1 *Let $C \subseteq k \subset \mathcal{S}$ be a perfect difference subfield and assume that the algebraic closure $\bar{k}$ of k also lies in $\mathcal{S}$. Let $A \in Gl(n, k)$ and consider the equation $\phi(Y) = AY$.*
Let N be such that $A = (A(1), A(2), A(3), \ldots)$, considered as an element of $Gl(n, \mathcal{S})$, satisfies $A(m) \in Gl(n, C)$ for $m \geq N$. Define $Z = (Z_{i,j}) \in Gl(n, \mathcal{S})$ by $Z(N) = id$ and $Z(m + 1) = A(m)Z(m)$ for $m \geq N$. Then:

1. *$\phi(Z) = AZ$ and $R = k[Z_{i,j}, \frac{1}{det(Z)}] \subset \mathcal{S}$ is the Picard-Vessiot ring associated to $\phi(Y) = AY$.*

2. *Every $Y \in \mathcal{S}^n$, solution of $\phi(Y) = AY$, is a C-linear combination of the columns of Z.*

Proof: 1. We consider the equation $\phi(Y) = AY$ first over $\bar{k}$. Let $H \subset Gl(n, C)$ be the difference Galois group of this equation over $\bar{k}$. Let P denote the Picard-Vessiot ring of the equation over $\bar{k}$. Since $\bar{k}$ is algebraically closed, $Spec(P)$ is a trivial H-torsor over $\bar{k}$. As in Proposition 1.21 this implies the existence of a $B \in Gl(n, \bar{k})$ such that the base change

$$Y = BX, \ \phi(X) = \phi(B)^{-1}ABX, \ D = \phi(B)^{-1}AB,$$

satisfies $D \in H(\bar{k})$. Moreover $P = \bar{k}[X_{i,j}, \frac{1}{det(X)}]/I$, where I is the ideal generated by the ideal of $H \subset Gl(n, C)$.

We view B and D as elements of $Gl(n, \mathcal{S})$. Take an integer M such that $D(m) \in H(C)$ for all $m \geq M$. Define $T \in Gl(n, \mathcal{S})$ by the formulas $T(M) = id$ and $T(m + 1) = D(m)T(m)$ for $m \geq M$.

Then $\phi(T) = DT$ and $T \in H(\mathcal{S})$, since $T(m) \in H(C)$ for all $m \geq M$. Write $T = (T_{i,j})$ and consider the morphism of $\bar{k}$-algebras $\bar{k}[X_{i,j}, \frac{1}{det(X)}] \to \bar{k}[T_{i,j}, \frac{1}{det(T)}] \subset \mathcal{S}$, given by $X_{i,j} \mapsto T_{i,j}$. This morphism is surjective and ϕ-equivariant. The kernel contains the ideal I since $T \in H(\mathcal{S})$. Since I is maximal among the ϕ-invariant ideals, we conclude that I is the kernel. Thus we find an isomorphism of difference rings $P \cong \bar{k}[T_{i,j}, \frac{1}{det(T)}] \subset \mathcal{S}$.

From the definition of Z it follows that $Z = BTE$, where E is some element of $Gl(n, C)$. Thus $\bar{k}[Z_{i,j}, \frac{1}{det(Z)}] \subset \mathcal{S}$ is the Picard-vessiot ring of the equation $\phi(Y) = AY$ over $\bar{k}$. The subring $R := k[Z_{i,j}, \frac{1}{det(Z)}]$ has no ϕ-invariant ideals $\neq 0, R$, since $R \otimes_k \bar{k}$ has no ϕ-invariant ideals $\neq 0, R \otimes_k \bar{k}$. The conclusion is that R is the Picard-Vessiot ring for the equation $\phi(Y) = AY$ over k.

2. Write $Y = Z\check{Y}$ with $\check{Y} \in \mathcal{S}^n$. Then clearly $\phi(\check{Y}) = \check{Y}$. Since the ring $\mathcal{S}$ has C as set of constants one finds that $\check{Y} \in C^n$ and the statement is proved.   ∎

The above Proposition implies that, for k as above, the ring $\mathcal{S}' = \{a \in S \mid a$ satisfies a linear homogeneous difference equation over $k\}$ is *the universal Picard-Vessiot ring for k.*

We now turn to reproving and extending results from [6, 7, 35, 55].

Definition 4.2 *A sequence* $\mathbf{y}$ *is an* interlacing *of the sequences* $\mathbf{a} = (a_0, a_1, \ldots)$, $\mathbf{b} = (b_0, b_1, \ldots), \ldots, \mathbf{e} = (e_0, e_1, \ldots)$ *if* $\mathbf{y} = (a_0, b_0, \ldots, e_0, a_1, b_1, \ldots, e_1, \ldots)$. *We will say that an element* $y \in \mathcal{S}$ *is an* interlacing *of elements* $a, b, \ldots, e \in \mathcal{S}$ *if there exist sequences in* $a, b, \ldots, e$ *whose interlacing lies in* y.

Proposition 4.3 *(c.f., [35], Theorem 1.2) Let* $C \subset k \subset \mathcal{S}$ *be a perfect difference field whose algebraic closure is also in* $\mathcal{S}$. *If* $u, v \in \mathcal{S}$ *satisfy linear difference equations over* k *and* $u \cdot v = 0$ *then* u *and* v *are the interlacing of sequences* $u_1, \ldots, u_t$ *and* $v_1, \ldots, v_t$ *such that for each* i *either* $u_i = 0$ *or* $v_i = 0$

Proof: Proposition 4.1 implies that u and v belong to a Picard-Vessiot extension R of k with $R \subset \mathcal{S}$. Let $R = e_0 R \oplus \ldots \oplus e_{t-1} R$. Since each e_i is idempotent and

$e_0 + \ldots + e_{t-1} = 1$, one sees that (after a possible renumbering) each e_i is the sequence whose j^{th} term is 1 if $j \equiv i (mod\ t)$ and 0 otherwise. Let $\tilde{u}_i = u e_i$ and $\tilde{v}_i = v e_i$. Denote the jth entry of $\tilde{u}_i$ (resp. $\tilde{v}_i$) by $u_{i,j}$ (resp. $v_{i,j}$) and define the sequences $u_i = (u_{i,i}, u_{i,i+t}, u_{i,i+2t}, \ldots)$ and $v_i = (v_{i,i}, v_{i,i+t}, v_{i,i+2t}, \ldots)$. Clearly u and v are the interlacing of sequences $u_1, \ldots, u_t$ and $v_1, \ldots, v_t$. Since each ring $e_i R$ is a domain we have that for each i, either $\tilde{u}_i$ or $\tilde{v}_i$ is 0. Therefore, for each i either $u_i = 0$ or $v_i = 0$.   ∎

Proposition 4.4 *(c.f., [7], Théorème 1) Let $C \subset k \subset S$ be a difference field whose algebraic closure is also in S. Let $u \in S$ satisfy a linear difference equation over k and assume that there exists a non-zero polynomial $P(Y) \in k[Y]$ such that $P(u) = 0$. Then u is the interlacing of sequences each of which lies in a finite algebraic difference field extension of k. If $k = C(z)$, then these elements lie in $C(z)$.*

Proof: We proceed as in the proof of Proposition 4.3. Corollary 4.1 implies that u belongs to a Picard-Vessiot extension of $k \subset S$. Let $R = e_0 R \oplus \ldots \oplus e_{t-1} R$. Since each e_i is idempotent and $e_0 + \ldots + e_{t-1} = 1$, one sees that (after a possible renumbering) each e_i is the sequence whose j^{th} term is 1 if $j \equiv 0 (mod\ t)$ and 0 otherwise. Let $\tilde{u}_i = u e_i$ and let $P_i(Y) = e_i \cdot P(Y)$. For each i, P_i is a nonzero polynomial satisfied by $\tilde{u}_i$. Since $e_i R$ is a finitely generated domain, $\tilde{u}_i$ will belong to a finitely generated algebraic extension of $e_i \cdot k \approx k$. If $k = C(z)$, Lemma 1.19 implies that $\tilde{u}_i \in e_i \cdot C(z)$. Let $\tilde{u}_i = e_i \cdot f_i$, $f_i \in C(z)$. Let $g_i(z) = f_i((z - i)/t)$. We then have that u is the interlacing of the sequences defined by the g_i.   ∎

The following result was conjectured in [7]. It is the analogue for difference equations of a result proven in [59] for differential equations.

Proposition 4.5 *(c.f., [6] Théorème 3, [7] Conjecture C_1, [35] Theorem 1.1) Let C be an algebraically closed field and $k = C$ or $C(z)$ (we allow this latter possibility only if C has characteristic zero). If $u \in S$ is invertible in S and u and $1/u$ satisfy linear difference equations over k then u is the interlacing of sequences u_i such that for each i, $\phi_0(u_i)/u_i \in k^*$.*

Proof: Proposition 4.1 implies that u and $1/u$ belong to a Picard-Vessiot extension $R \subset S$ of k. Writing $R = R_0 \oplus \ldots \oplus R_{t-1}$ as in Corollary 1.16 implies that we may write u as an interlacing of sequences $u_0, \ldots, u_{t-1}$, each of which has the property that u_i and $1/u_i$ belong to a Picard-Vessiot domain R_i over k (where the automorphism is now ϕ^t). Since k is cohomologically trivial, Corollary 1.18 implies that each R_i is of the form $k[G]$, where G is a connected group (since R_i is a domain). Fix some R_i, say R_0. We now invoke a theorem of Rosenlicht [38, 56, 59] which states: *Let G be a connected linear algebraic group defined over an algebraically closed field $\bar{k}$ and y is a regular functions on G with $1/y$*

also a regular function. Then y is a $\overline{k}$ multiple of a character. We apply this result to R_0. We can conclude that if $\psi \in G(C)$ then $\psi(u_i) = a_\psi u_i$ for some $a_\psi \in \overline{k} \cap R = k$.

We shall now show that $a_\psi \in C_R$. Since u_0 satisfies a linear difference equation over R_0, let $L(u_0) = a_n \phi^{tn}(u_0) + a_m \phi^{tm}(u_0) + \ldots + a_0 u_0 = 0$ be such an equation of lowest order. We assume that $a_n a_m \neq 0$. Applying ψ we have

$$
\begin{aligned}
L(\psi(u_0)) &= a_n \phi^{tn}(\psi(u_0)) + a_m \phi^{tm}(\psi(u_0)) + \ldots + a_0 \psi(u_0) \\
&= a_n \phi^{tn}(a_\psi u_0) + a_m \phi^{tm}(a_\psi u_0) + \ldots + a_0 a_\psi u_0 \\
&= a_n \phi^{tn}(a_\psi) \phi^{tn}(u_0) + a_m \phi^{tm}(a_\psi) \phi^{tm}(u_0) + \ldots + a_0 a_\psi u_0 \\
&= 0
\end{aligned}
$$

By minimality, we must have that $\phi^{tn}(a_\psi) = \phi^{tm}(a_\psi)$ and so $\phi^{t(n-m)}(a_\psi) = a_\psi$. Since a_ψ is left fixed by some power of ϕ, it is algebraic over C and so must lie in C.

Since $a_\psi \in C_R$ for all $\psi \in Gal(R/k)$ we have that $\phi^t(u_0)/u_0$ is left fixed by all $\psi \in Gal(R/k)$. Therefore $\phi^t(u_0)/u_0 \in k$. A similar argument applies to all the u_i.   ∎

We note that Proposition 4.3 was proven in [35] for sequences u, v of elements in a field C satisfying a linear difference equation over C, i.e., for *linearly recursive sequences*. Their proof does not extend to the case treated above. Proposition 4.5 is proven for linearly recursive sequences in [6] and [35]. In [7], this result is conjectured (Conjecture C_1) for sequences that satisfy linear difference equations over $C(x)$, i.e., for *differentially finite sequences*, and proven when u satisfies a second order linear difference equation. We close this section by settling the other conjectures of this latter paper. A linearly recursive sequence $a \in \mathcal{S}$ is said to be an *exponential sum* if there exist $c_i, \lambda_i \in C$ such that the nth term of a is of the form $\sum_{i=1}^m c_i \lambda_i^n$. Note that this is equivalent to requiring that a satisfies a homogeneous linear difference equation over C whose indicial polynomial has no repeated roots.

Proposition 4.6 *(c.f., [7] Conjecture C_2) Let C be an algebraically closed field of characteristic zero. Let $a_n, \ldots, a_0 \in \mathcal{S}$ satisfy linear difference equations over C and assume that at least one of these elements is invertible in $\mathcal{S}$. If $u \in \mathcal{S}$ satisfies a linear difference equation over $C(z)$ and satisfies*

$$
a_n u^n + \ldots + a_0 = 0
$$

then there exists a nonzero polynomial $P(z) \in C[z]$ such that $P \cdot u$ satisfies a linear difference equation over C. Furthermore, if the a_i are exponential sums, then we can conclude that u is an exponential sum.

The second part of Proposition 4.6 is conjectured (Conjecture C_2) in [7] and proven when u satisfies an equation of the form $u^r - a_0 = 0$. The first part of Proposition 4.6 is conjectured in the final *Remarques* of this latter paper. Before we prove the above proposition, we gather some facts concerning elements z of a difference field having the property that $\phi(z)/z$ lies in a subfield. Recall that an algebraic group in $GL(n, C)$ is a torus if it is connected and diagonalizable.

Proposition 4.7 *Let $k \subset K$ be difference fields with the same algebraically closed field of constants C.*

1. If K is a Picard-Vessiot extension of k, then $G(K/k)$ is conjugate to a subgroup of a torus if and only if there exist $z_1, \ldots, z_r$ in K^* such that $\phi(z_i)/z_i \in k$ for $i = 1, \ldots, r$ and $K = k(z_1, \ldots, z_r)$.

2. Let $z_1, \ldots, z_r \in K^*$ and assume that $\phi(z_i)/z_i \in k$ for $i = 1, \ldots, r$. If z_1 is algebraic over $k(z_2, \ldots, z_r)$, then there exist $n_1 \neq 0, n_2, \ldots n_r \in \mathbf{Z}$ such that $\prod_{i=1}^r z_i^{n_i} \in k$

Proof: 1. Let $K = k(Y)$, where $\phi(Y) = AY$, $A \in GL(n, C)$. If $G(K/k)$ is conjugate to subgroup of a torus, then there exists a basis $\{y_1, \ldots, y_n\}$ of the solution space of $\phi(Y) = AY$ such that for any $g \in G(K/k)$, $g(y_i) = c_i^g y_i$ for constants $c_i^g \in C$. If we write $y_i = (y_{i1}, \ldots, y_{in})$ then one sees that for $y_{ij} \neq 0$, $\phi(y_{ij})/y_{ij}$ is left fixed by all elements of $G(K/k)$ and so must lie in k. Conversely, assume that there exist $z_1, \ldots, z_r$ in K such that $\phi(z_i)/z_i = u_i \in k$ for $i = 1, \ldots, r$ and $K = k(z_1, \ldots, z_r)$. Let $A = diag(u_1, \ldots, u_r)$ and $Y = diag(z_1, \ldots, z_r)$. We then have that K is the Picard-Vessiot extension of k corresponding to $\phi(Y) = AY$.

2. If $z_2, \ldots, z_n$ are algebraic over k, then z_1 is also algebraic over k. Clearly $k(z_1)$ is a Picard-Vessiot extension of k whose Galois group is a finite subgroup of C^* and so is cyclic of some order, say n_1. Therefore, $z_1^{n_1}$ is left fixed by the Galois group and so must lie in k. We can therefore now assume that $z_2, \ldots, z_r$ are algebraically independent. We have that $k(z_1, \ldots, z_r)$ is a Picard-Vessiot extension of k whose Galois group is a subgroup of $(C^*)^r$. Since the transcendence degree of $k(z_1, \ldots, z_r)$ over k equals the dimension of the Galois group, we must have that this group is a proper subgroup of $(C^*)^r$. Therefore there exist $n_1, \ldots, n_r \in \mathbf{Z}$, not all zero, such that any element $g = diag(d_1, \ldots, d_r) \in G(k(z_1, \ldots, z_r)/k)$ satisfies $\prod_{i=1}^r d_i^{n_i} = 1$. This implies that $\prod_{i=1}^r z_i^{n_i}$ is left fixed by the Galois group and so must be in k. If $n_1 = 0$ then $z_2, \ldots, z_r$ would be algebraically dependent, a contradiction. ∎

Proof of Proposition 4.6: Since $a_n, \ldots, a_0$ satisfy linear difference equations over C, they belong to a Picard-Vessiot extension of C generated by elements $y_1, \ldots, y_m$ and z where $\phi(y_i)/y_i \in C$. Let $R \subset S$ be a Picard-Vessiot extension of $C(z)$ containing $y_1, \ldots, y_m$ and u. Let $R = R_0 \oplus \ldots \oplus R_{t-1}$ be the decomposition into the direct sum of Picard-Vessiot domains with respect to ϕ^t and let π_i denote the projection onto R_i. The same hypotheses apply to

$\pi_i(a_n), \ldots, \pi_i(a_0), \pi_i(u)$ with respect to $\pi_i(C(z))$ and ϕ^t. If we can find polynomials P_i such that $\pi_i(P_i) \cdot \pi_i(u)$ satisfies a linear difference equation over $C(z)$ then $P = \prod P_i$ satisfies the conclusion of the Theorem (note that if v satisfies a linear difference equation over C and Q is any polynomial, then $Q \cdot v$ also satisfies a linear difference equation over C).

Therefore, we will assume from the start that R is a Picard-Vessiot domain. Let $k = C(z)$ and $K =$ the quotient field of R. Let $F = k(y_1, \ldots, y_m)$ and $E =$ the algebraic closure of F in K. Both F and E are left invariant by the Galois group of K over k. Note that since K is finitely generated over k, E is a finite normal algebraic extension of F. We have an exact sequence of algebraic groups

$$1 \longrightarrow G(E/F) \longrightarrow G(E/k) \longrightarrow G(F/k) \longrightarrow 1$$

Applying Proposition 4.7, we see that $G(F/k)$ is a subgroup of a torus and so all of its elements are semisimple. Since $G(E/F)$ is finite all of its elements are semisimple. Therefore all elements of $G(E/k)$ are semisimple, so this latter group is a finite extension of a torus T. The fixed field of T will be a finite algebraic difference extension of k and so must equal k. Therefore, the Galois group $G(E/k)$ is a torus T. Proposition 4.7.1 implies that $E = k(z_1, \ldots, z_m)$ with $\phi(z_i)/z_i \in k$. Since k is cohomologically trivial, we have that $E = k \otimes C(T) = k(T)$. The elements z of E such that the T orbit of z spans a finite dimensional C−vector space are the elements of $k[T] = k[z_1, \ldots, z_m, z_1^{-1}, \ldots, z_m^{-1}]$. Therefore, u is a polynomial in $z_1, \ldots, z_m, z_1^{-1}, \ldots, z_m^{-1}$, with coefficients in $C(z)$. Proposition 4.7.2 implies that for each i, there are integers n_{ij} such that

$$z_i = (f_i \prod_j y_j^{n_{ij}})^{\frac{1}{n_{i0}}}$$

where $f_i \in C(z)$. This formula implies that $f_i^{\frac{1}{n_{i0}}}$ satisfies a linear difference equation over $C(z)$. Since it is also algebraic over $C(z)$, it must lie in $C(z)$. Let $N = lcm_i\{n_{i0}\}$. We then have that any polynomial in $z_1, \ldots, z_m, z_1^{-1}, \ldots, z_m^{-1}$, with coefficients in $C(z)$ can be written as a polynomial in $y_1^{\frac{1}{N}}, \ldots, y_m^{\frac{1}{N}}, y_1^{\frac{-1}{N}}, \ldots, y_m^{\frac{-1}{N}}$ with coefficients in $C(z)$ and therefore u is of this form. Clearing denominators, we see that there exists a polynomial $P \in C(z)$ such that $P \cdot u$ is a polynomial in $y_1^{\frac{1}{N}}, \ldots, y_m^{\frac{1}{N}}, y_1^{\frac{-1}{N}}, \ldots, y_m^{\frac{-1}{N}}$ and z with coefficients in C and therefore satisfies a linear difference equation with coefficients in C.

Now assume that the a_i are exponential sums. This implies that they belong to a Picard-Vessiot extension of C generated by elements y_i with $\phi(y_i)/y_i \in C$, i.e., z is not needed. We may assume that the y_i are algebraically independent over C. From the above, we know that u is algebraic over $C(y_1, \ldots, y_m)$ and can be expressed as a polynomial in $y_1^{\frac{1}{N}}, \ldots, y_m^{\frac{1}{N}}$, $y_1^{\frac{-1}{N}}, \ldots, y_m^{\frac{-1}{N}}$ with coefficients that are in $C(z)$. Let $w_i = y_i^{\frac{1}{N}}$. To prove the final claim of the Proposition, it will therefore suffice to show that $w_1, \ldots, w_m$ and z are algebraically independent over C. If not then $L = C(w_1, \ldots, w_m, z)$ is a finite extension of $C(w_1, \ldots, w_m)$ and, as we have seen above, this implies that the Galois group G of L over C must be a torus. One sees that for any automorphism $g \in G$, we have that $g(z) = z + c_g$ with $c_g \in C$ and $g(1) = 1$. The action of G on the C-space V spanned by z and 1 is rational and the above calculation shows that each element of G gets mapped to a unipotent element of $GL(V)$. Since each element of G is semisimple, we have that G must act trivially on V. This implies that z is left fixed by G and so must lie in C, a contradiction.

∎

Chapter 5

An excursion in positive characteristic

5.1   Generalities

A difference equation $y(z + 1) = A\ y(z)$ where $A \in Gl(n, \mathbf{Q}(z))$ can be reduced modulo p for almost all primes p. It turns out that the reduced equation, which has the form $y(z + 1) = B\ y(z)$ where $B \in Gl(n, \mathbf{F}_p(z))$, is easier to solve. Further the solutions in characteristic p should give information about the original equation.

In what follows we will not try to give a general theory for difference equations in characteristic p. We restrict ourselves to a field $k(z)$ with $k = \mathbf{F}_p$, $\mathbf{F}_q$ or $\overline{\mathbf{F}}_p$ with the automorphism ϕ given by $\phi(z) = z + 1$ and ϕ is the identity on k. This leads to a ring of difference operators $\mathcal{D} = k(z)[\Phi, \Phi^{-1}]$ with its structure given by $\Phi f = \phi(f)\Phi$.

As usual, linear difference equations $y(z + 1) = A\ y(z)$ can be translated into left $\mathcal{D}$-modules which are finite dimensional over $k(z)$. The center Z of $\mathcal{D}$ is equal to $k(z^p - z)[\Phi^p, \Phi^{-p}]$. For the choice $k = \overline{\mathbf{F}}_p$ there is an equivalence between the category of Z-modules of finite dimension over $k(z^p - z)$ and the category of left $\mathcal{D}$-modules of finite dimension over $k(z)$. This equivalence respects tensor products. The consequences of this are:

- A simple classification of difference modules over $k(z)$.

- The difference Galois group of a difference module M is the (commutative) algebraic group over $k(z^p - z)$ generated by the "p-curvature of M", i.e. the action of Φ^p on M as vector space over $k(z^p - z)$. Over the algebraic closure of $k(z^p - z)$, this group is isomorphic to a product of a finite cyclic group and an algebraic torus.

The naive translation of Grothendieck's conjecture on differential equations over $\mathbf{Q}(z)$ for the case of difference equations reads:

The difference equation $y(z + 1) = A\, y(z)$ over $\mathbf{Q}(z)$ is trivial if and only if for almost all p the p-curvature is 1.

It is a surprise that this statement is false. At present we do not know what the "correct" translation of Grothendieck's conjecture would be.

We start with the investigation of the skew ring $\mathcal{D}$. In the sequel we will assume that the field k is equal to $\overline{\mathbf{F}_p}$ and we remark that many results are valid without this assumption. The results and methods in this section are very similar to the situation of differential equations in characteristic $p > 0$ (See [50]). Therefore we have been brief in the proofs of the statements.

Proposition 5.1

1. The center Z of $\mathcal{D}$ is equal to $k(z^p - z)[\Phi^p, \Phi^{-p}]$.

2. $\mathcal{D}$ is a free Z-module of rank p^2.

3. Let $\underline{m}$ denote a maximal ideal of Z with residue field $L = Z/\underline{m}$. Then $\mathcal{D}/\underline{m}\mathcal{D} = L \otimes_Z \mathcal{D}$ is isomorphic to $M(p, L)$, the algebra of $p \times p$ matrices over L.

4. For any power $\underline{m}^n$ of the maximal ideal $\underline{m}$, the algebra $\mathcal{D}/\underline{m}^n\mathcal{D} = Z/\underline{m}^n \otimes_Z \mathcal{D}$ is isomorphic to $M(p, Z/\underline{m}^n)$. Let $\hat{Z}_{\underline{m}}$ denote the completion of Z with respect to the ideal $\underline{m}$, then $\hat{Z}_{\underline{m}} \otimes_Z \mathcal{D} \cong M(p, \hat{Z}_{\underline{m}})$.

5. Let $\hat{Z}$ denote the completion of Z with respect to the set of all non zero ideals, then $\hat{Z} \otimes_Z \mathcal{D} \cong M(p, \hat{Z})$.

Proof. 1. Clearly $\tilde{Z} := k(z^p - z)[\Phi^p, \Phi^{-p}]$ is contained in Z. Any element f in $\mathcal{D}$ has a unique presentation $f = \sum_{0 \le n, m < p} a(n, m) z^n \Phi^m$ with all $a(n, m) \in \tilde{Z}$. Suppose that $f \in Z$, then $0 = \Phi f \Phi^{-1} - f = \sum a(n, m)\{(z + 1)^n - z^n)\}\Phi^m$ implies that $a(n, m) = 0$ for $n \ne 0$. Further $0 = fz - zf = \sum_m a(0, m) m \Phi^m$ implies that $a(0, m) = 0$ for $m > 0$. This proves $Z = \tilde{Z}$.

Statement 2. is obvious.

3. We show first that $L \otimes_Z \mathcal{D}$ is a central simple algebra over L. As in 1. one proves that the center of $L \otimes_Z \mathcal{D}$ is L. Suppose that I is a two-sided ideal of $L \otimes_Z \mathcal{D}$, containing a non zero element $f = \sum_{0 \leq n, m < p} a(n, m) z^n \Phi^m$ with all $a(n, m) \in L$. If the degree of f with respect to z is not zero then $\Phi f \Phi^{-1} - f$ is a non zero element in I of lower degree. Hence I contains some non zero element, again called f, with degree 0 in z. After dividing f by a power of Φ we may suppose that f has the form $f = \sum_m a(0, m) \Phi^m$ with $a(0, 0) \neq 0$. If f has more than one term then $(fz - zf)\Phi^{-1}$ is a similar expression with less terms. Finally we conclude that I contains a non zero element in L and thus $I = L \otimes_Z \mathcal{D}$. This shows that $L \otimes_Z \mathcal{D}$ is a central simple algebra over L of degree p^2. Therefore $L \otimes_Z \mathcal{D}$ is isomorphic to a ring of matrices over a division ring having L as its center. Our assumption that k is algebraically closed implies that $k(z^p - z)$ and also L are C_1 fields. Therefore there are no division rings with center L of finite dimension over L. Since p is prime, the only possibility for $L \otimes_Z \mathcal{D}$ is $M(p, L)$.

4. The isomorphism $Z/\underline{m} \otimes_Z \mathcal{D} \to M(p, Z/\underline{m})$ implies that there is a left $\mathcal{D}$-module M such that M is a vector space over $Z/\underline{m}$ of dimension p. One can consider M as a module over $Z/\underline{m}[z]$. This module is free with one generator e. The action of Φ on $M = Z/\underline{m}[z]e$ is given by $\Phi(e) = ve$ for a certain invertible element $v \in Z/\underline{m}[z]$. Let t denote the image of Φ^p in $Z/\underline{m}$. Then $v = v(z)$ must satisfy the equality $v(z + p - 1) \ldots v(z + 1)v(z) = t$.
We will use this to give the free $\hat{Z}_{\underline{m}}[z]$-module of rank one $N := \hat{Z}_{\underline{m}}[z]e$ a structure of a left $\hat{Z}_{\underline{m}} \otimes_Z \mathcal{D}$-module. It suffices to define $\Phi(e) = ue$ for a suitable invertible element u of $\hat{Z}_{\underline{m}}[z]$. The only condition on u is $u(z+p-1) \ldots u(z+1)u(z) = T$, where T denotes the image of Φ^p in $\hat{Z}_{\underline{m}}$. We will in fact produce a u such that $u \bmod \underline{m}$ is equal to v above.

Suppose that u is found. Then $\hat{Z}_{\underline{m}} \otimes_Z \mathcal{D}$ acts on N and we find a morphism of algebras $f : \hat{Z}_{\underline{m}} \otimes_Z \mathcal{D} \to End(N)$, where $End(N)$ denotes the algebra of $\hat{Z}_{\underline{m}}$-linear endomorphisms of N. This last algebra can be identified with $M(p, \hat{Z}_{\underline{m}})$. Consider f as $\hat{Z}_{\underline{m}}$-linear map between two free $\hat{Z}_{\underline{m}}$-modules of rank p^2. By construction, the reduction modulo $\underline{m}$ of f coincides with the bijection $Z/\underline{m} \otimes_Z \mathcal{D} \to M(p, Z/\underline{m})$. Hence f itself is a bijection. By reducing f modulo powers of $\underline{m}$ one finds the other statements of 4.

For the construction of the invertible element u of $\hat{Z}_{\underline{m}}$ is suffices to prove the following assertion:

Let $k \geq 1$. Suppose that there is an invertible element u_k in $\hat{Z}_{\underline{m}}[z]$ such that $u_k(z + p - 1) \ldots u_k(z + 1)u_k(z)$ is congruent to T modulo $\underline{m}^k$. Then there is an element u_{k+1} of the form $u_k(1 + F^k b)$, with F a generator of the ideal $\underline{m}$ and $b \in \hat{Z}_{\underline{m}}[z]$, such that $u_{k+1}(z + p - 1) \ldots u_{k+1}(z + 1)u_{k+1}(z)$ is congruent to T modulo $\underline{m}^{k+1}$.

A small calculation shows that the element b should satisfy $b(z) + b(z+1) +$
$\ldots b(z + p - 1)$ is congruent modulo $\underline{m}$ with the expression

$$a := \frac{T - u_k(z + p - 1) \ldots u_k(z + 1)u_k(z)}{u_k(z + p - 1) \ldots u_k(z + 1)u_k(z)F^k}.$$

The element a is seen to lie in $\hat{Z}_{\underline{m}}$. The choice $b = \frac{az^{p-1}}{p-1}$ has the required property as is easily verified.

5. Since Z is a ring of Laurent polynomials over a field it follows that $\hat{Z} = \prod_{\underline{m}} \hat{Z}_{\underline{m}}$ where the product is taken over the set of all maximal ideals of Z. Then 5. follows at once from 4.   ∎

Proposition 5.2 *The category of Z-modules of finite dimension over $k(z^p - z)$ is equivalent to the category of the left $\mathcal{D}$-modules which are of finite dimension over $k(z)$. This isomorphism preserves tensor products.*

Proof. The category of $\hat{Z}$-modules of finite dimension over $k(z^p - z)$ coincides with the category of Z-modules of finite dimension over $k(z^p - z)$. The same statement holds for $\hat{\mathcal{D}} := \hat{Z} \otimes_Z \mathcal{D}$. We use part 5 of Proposition 5.1 and combine this with Morita's equivalence.
This equivalence can be formulated as follows (See [54], Proposition 17, p.19):

Let A be any (unitary) ring. The categories of left A-modules and left $M(n, A)$-modules are equivalent. The equivalence is given by $M \mapsto M^n$, with the obvious left action of $M(n, A)$ on M^n.

This concludes the construction of the equivalence $\mathcal{F} : N \mapsto M$ between the two categories. A more explicit way is to define $\mathcal{F}(N)$ as $k(z) \otimes_{k(z^p - z)} N$. Its structure as a $k(z)[\Phi^p, \Phi^{-p}]$-module is clear. The construction above leads to a compatible Φ-action on M.

Finally we have to say something about tensor products. Let L be any field and let $FMod_{L[T,T^{-1}]}$ denote the category of the modules over $L[T, T^{-1}]$, which have finite dimension over L. This abelian category is given a "tensor structure" by defining the tensor product of two modules M and N as $M \otimes_L N$ with the operation of T given by the formula $T(m \otimes n) = (Tm) \otimes (Tn)$. This explains the tensor product for the first category of Proposition 5.2. The tensor product for difference modules M and N over $k(z)$ is defined in a similar way, namely: The tensor product is $M \otimes_{k(z)} N$ with the operation of Φ given by $\Phi(m \otimes n) = (\Phi m) \otimes (\Phi n)$. It is not difficult to show, using the description $\mathcal{F}(N) = k(z) \otimes_{k(z^p - z)} N$, that $\mathcal{F}$ respects tensor products.   ∎

Corollary 5.3 *For every maximal ideal $\underline{m}$ in $\mathbf{Z}$ and every $n \geq 1$ one defines the indecomposable left $\mathcal{D}$-module $I(\underline{m}^n)$ to be $\mathcal{F}(Z/\underline{m}^n) = (Z/\underline{m}^n) \otimes_{k(z^p - z)} k(z)$. Every left $\mathcal{D}$-module M of finite dimension over $k(z)$ is isomorphic to a direct sum*

$$\sum_{\underline{m},n} I(\underline{m}^n)^{e(\underline{m},n)}.$$

The numbers $e(\underline{m}, n) \geq 0$ are uniquely determined by M.

Proof. The indecomposable Z-modules of finite dimension over $k(z^p - z)$ are the $Z/\underline{m}^n$. Hence Corollary 5.3 follows at once from Proposition 5.2. ∎

Definition 5.4 *The p-curvature of a difference module M over $k(z)$ is the $k(z^p - z)$-linear action of Φ^p on the module N with $\mathcal{F}(N) = M$.*

By construction the $k(z^p - z)$-linear Φ^p on N extends to the $k(z)$-linear Φ^p on $M = k(z) \otimes_{k(z^p - z)} N$. We will also call Φ^p on M the p-curvature when no confusion occurs. The name is copied from a similar situation for differential equations in characteristic p. Let M be a difference module over $k(z)$ which is represented by an equation in matrix form $y(z + 1) = A\, y(z)$, where $A = A(z)$ is an invertible matrix with coefficients in $k(z)$. Then the p-curvature (as a $k(z)$-linear map on M) has the matrix $A(z + p - 1) \ldots A(z + 1)A(z)$.

5.2   Modules over $K[T, T^{-1}]$

For the moment K is any field. The modules M over $K[T, T^{-1}]$ are supposed to be finite dimensional vector spaces over K. In other words, a module is the same thing as a vector space M over K of finite dimension together with an invertible linear map (i.e. the action of T on M). The category of all those modules is denoted by $FMod_{K[T,T^{-1}]}$. It is in an obvious way an abelian category. The structure of the tensor category is defined in the proof of Proposition 5.2. The functor $\omega : FMod_{K[T,T^{-1}]} \to Vect_K$, where $Vect_K$ denotes the category of finite dimensional vector spaces over K, is the forgetful functor $\omega(M) = M$ (i.e. one forgets on M the action of T). One can verify that ω is a fibre functor. This makes $FMod_{K[T,T^{-1}]}$ into a neutral Tannakian category. Associated to this is an affine group scheme G over K. This affine group scheme represents the functor $R \mapsto Aut^{\otimes}(\omega)(R)$ defined on K-algebras.

For a fixed object $M \in FMod_{K[T,T^{-1}]}$ one can consider the tensor subcategory $\{\{M\}\}$ of $FMod_{K[T,T^{-1}]}$ generated by M. This is the full subcategory whose objects are the subquotients of the tensor products

$M \otimes M \otimes \ldots \otimes M \otimes M^* \otimes \ldots \otimes M^*$ (as usual M^* denotes the dual of M). The restriction of ω to $\{\{M\}\}$ is again a fibre functor. The affine linear group associated to this $\{\{M\}\}$ is denoted by G_M.

It is an exercise to show that G_M is the smallest linear algebraic subgroup of $Gl(M)$ which contains the action of T on M.

We note that for an algebraically closed field K of characteristic 0, this group G_M is a direct product of a torus, a finite cyclic group and possibly a $\mathbf{G}_a$. The $\mathbf{G}_a$ is present precisely when T on M is not semi-simple.

If the characteristic of K is $p > 0$ then G_M is a direct product of a finite cyclic group and a torus. Indeed, suppose again that K is algebraically closed and let $T = T_{ss}T_u$ be the decomposition of the action of T on M into a semi-simple and a unipotent part. The group G_M is generated as an algebraic group by T_{ss} and T_u. The group generated by T_{ss} is easily seen to be a product of a torus and a cyclic group with order prime to p. The group generated by T_u is finite of order p^n where n is minimal such that the nilpotent matrix $T_u - 1$ satisfies $(T_u - 1)^{p^n} = 0$.

5.3 Difference Galois groups

For difference equations over the field $k(z)$ one would like to have a suitable theory of Picard-Vessiot extensions. The difference Galois group of an equation would then be the group of automorphisms of the Picard-Vessiot ring of the equation.

It is not excluded that such a Picard-Vessiot theory exists, in the sequel however we will use the theory of Tannakian categories for the definition and study of the difference Galois group. The main idea is to compare difference modules over $k(z)$ with modules over $Z = K[T, T^{-1}]$, where $K = k(z^p - z)$.

Let M be a difference module over $k(z)$ and let N be the Z-module with $\mathcal{F}(N) = M$. The category of difference modules $\{\{M\}\}$ is defined in a way similar to Chapter 1.4. The functor $\mathcal{F}$ induces an equivalence $\{\{N\}\} \to \{\{M\}\}$ of tensor categories. Hence $\{\{M\}\}$ is also a neutral Tannakian category with fibre functor

$$\{\{M\}\} \xrightarrow{\mathcal{F}^{-1}} \{\{N\}\} \xrightarrow{\omega} Vect_{k(z_p - z)}.$$

The *difference Galois group of M* is defined as the linear algebraic group over K associated to this fibre functor. This group is of course isomorphic to the group G_N of N defined in Section 5.2. Thus we find the following properties:

1. *The difference Galois group of a difference equation over $k(z)$ is the algebraic group over $k(z^p - z)$ generated by the p-curvature Φ^p.*

2. *The difference Galois group is a direct product of a finite cyclic group and a torus.*

5.4   Comparing characteristic 0 and p

We start by considering the following example of an order one equation:

$$y(z+1) = \frac{z+1/2}{z}y(z) \text{ over } \mathbf{Q}(z).$$

The only algebraic solution of this equation is 0. However, for every prime $p > 2$ the p-curvature is

$$\frac{z+p-1+1/2}{z+p-1} \cdots \frac{z+1+1/2}{z+1} \frac{z+1/2}{z} = 1.$$

This is clearly a counter example to the naive translation of Grothendieck's conjecture for difference equations. How to explain this? A somewhat trivial explanation is that the reduction of a rational number modulo a prime is an integer modulo a prime. This leads to the following result.

Lemma 5.5 *Consider the equation $y(z + 1) = ay(z)$ with $a \in \mathbf{Q}(z)^*$. The following statements are equivalent:*

1. *For almost all p the p-curvature is 1.*

2. *a has the form $\frac{b(z+\lambda)}{b(z)}$ where $b \in \mathbf{Q}(z)$ and $\lambda \in \mathbf{Q}$.*

3. *$a(\infty) = 1$ and for every algebraic number α the restriction of the divisor of a to any $\mathbf{Q}$-orbit $\alpha + \mathbf{Q} \subset \overline{\mathbf{Q}}$ has degree 0.*

Proof. Clearly 2. implies 1. and 2. implies 3. To see that 3. implies 2. note that there exists an integer q such that the restriction of the divisor of a to a $\mathbf{Q}$-orbit actually lies in a set of the form $\alpha + \frac{1}{q}\mathbf{Z}$. Let $b(z) = a(\frac{z}{q})$. The restriction of the divisor of B to a $\mathbf{Q}$-orbit is now actually a $\mathbf{Z}$-obit. We can therefore apply Proposition 2.1 to conclude that $b(z) = \frac{g(z+1)}{g(z)}$. Therefore $a(z) = \frac{g(qz+1)}{g(qz)}$. Dividing the numerator and denominator of this quotient by a suitable power of q, we have that $a(z) = \frac{h(z+1/q)}{h(z)}$ for a suitable h.

Now suppose that 1. holds. Then a cannot have a pole or zero at ∞ and the value $a(\infty) \in \mathbf{Q}^*$ has the property that for almost all p, one has $a(\infty)^p \equiv 1$

mod p. This implies that $a(\infty) = 1$. Write $a = a_1 \ldots a_s$, where the support of the divisor of each a_i lies in the union of $\{\infty\}$ with the $\mathbf{Q}$-orbit of an algebraic number α_i. The α_i are supposed to be distinct modulo $\mathbf{Q}$. Each term a_i has the form $(z - \alpha_i)^{n_i} b_i$, where $n_i \in \mathbf{Z}$ and where b_i satisfies condition 3.

The primes p that we consider are now prime ideals $\neq 0$ in the ring of integers of the field $\mathbf{Q}(\alpha_1, \ldots, \alpha_s)$. For almost all p the reductions modulo p of the α_i are distinct modulo $\mathbf{F}_p$. The p-curvature for almost all primes is

$$(z^p - (\alpha_1)^{p-1})^{n_1} \ldots (z^p - (\alpha_s)^{p-1})^{n_s} \equiv 1 \bmod p.$$

There is no cancellation between the various factors. Hence all $n_i = 0$ and a satisfies 3. ∎

Another interesting example is

$$y(z + 1) = \frac{z^2 - 2}{z^2} y(z)$$

We claim that for $p > 2$ the p-curvature is 0 if and only 2 is a square modulo p. Using the formula $\prod_{i=0,\ldots,p-1}(z + i - \alpha) \equiv z^p - (\alpha)^{p-1} z \bmod p$, one finds that the p-curvature of the expression above is $\frac{(z^p - 2^{(p-1)/2} z)^2}{(z^p - z)^2} \bmod p$. The p-curvature is 1 if and only if $2^{(p-1)/2} \equiv 1 \bmod p$. This proves the statement.

Finally, we give a general example where p-curvature 1 occurs for almost all p. Suppose that the equation $y(z + 1) = A\, y(z)$ with $A \in Gl(n, \mathbf{Q}(z))$ has a formal fundamental matrix $F(z) \in Gl(n, \mathbf{Q}((z^{-1})))$). Suppose moreover that a reduction modulo p of F exists for almost all p. This is the case when there are only finitely many primes in the denominators of the coefficients of the entries of F. Then, for almost all primes p, one has the equality $A(z + p - 1)\ldots A(z + 1)A(z) = F(z + p)F(z)^{-1} \equiv 1 \bmod p$.

These examples give the impression that the p-curvatures of an equation over $\mathbf{Q}(z)$ contain arithmetical information about the fundamental solution of the equation.

Chapter 6

Difference modules over $\mathcal{P}$

6.1 Classification of difference modules over $\mathcal{P}$

We recall that $\mathcal{P}$ is the algebraic closure of $\mathbf{C}((t))$ where $t = z^{-1}$ and $\phi(t^{\frac{1}{m}}) = t^{\frac{1}{m}}(1+t)^{-\frac{1}{m}}$. A one dimensional difference module over $\mathcal{P}$ is given as $\Phi e_f = f e_f$ with $f \in \mathcal{P}^*$. Let us call this module $\mathcal{E}(f)$. Clearly $\mathcal{E}(f_1)$ and $\mathcal{E}(f_2)$ are isomorphic if and only if $f_1 f_2^{-1}$ has the form $\frac{\phi(a)}{a}$ for some $a \in \mathcal{P}^*$. Write a as $a_0 t^\lambda (1 + \sum a_\mu t^\mu)$ with $a_0 \in \mathbf{C}^*$, $\lambda \in \mathbf{Q}$, $a_\mu \in \mathbf{C}$, $\mu > 0$, $\mu \in \frac{1}{m}\mathbf{Z}$ for some $m \geq 1$ (depending on a). A calculation of $\frac{\phi(a)}{a}$ yields that the subgroup U of $\mathcal{P}^*$ consisting of all elements $\frac{\phi(a)}{a}$ is equal to

$$U := \{(1 + \lambda t + \sum_{\mu > 1} b_\mu t^\mu)| \; \lambda \in \mathbf{Q}, \; b_\mu \in \mathbf{C}, \; \mu \in \frac{1}{m}\mathbf{Z} \text{ for some } m \geq 1\}$$

For the group $\mathcal{P}^*/U$ one can take various subgroups $\mathcal{G}$ of $\mathcal{P}^*$ as set of representatives. We will take $\mathcal{G}$ as follows:

$$\mathcal{G} = \{t^\lambda \, c \, (1 + t)^{a_0} exp(\phi(q) - q)| \lambda \in \mathbf{Q}, \; c \in \mathbf{C}^*, \; q = \sum_{0 < \mu < 1} a_\mu z^\mu\}$$

where $\mu \in \mathbf{Q}$, $a_\mu \in \mathbf{C}$ and where the sum $\sum_{0 < \mu < 1} a_\mu z^\mu$ is supposed to be a finite sum and where a_0 is supposed to lie in a $\mathbf{Q}$-linear subspace L of $\mathbf{C}$ satisfying $L \oplus \mathbf{Q} = \mathbf{C}$.

Any 1-dimensional difference module over $\mathcal{P}$ is isomorphic to a unique $\mathcal{E}(g)$ with $g \in \mathcal{G}$. Furthermore there is a natural isomorphism $\mathcal{E}(g_1) \otimes \mathcal{E}(g_2) \to \mathcal{E}(g_1 g_2)$, given by $e_{g_1} \otimes e_{g_2} \mapsto e_{g_1 g_2}$. We note that our choice for $\mathcal{G}$ is inspired by the interpretation of the formal solution of the difference equation as (the inverse of) the multivalued function $\Gamma^\lambda c^z z^{a_0} exp(q)$.

A difference module M is called *unipotent* if there exists a sequence of submodules $0 = M_0 \subset M_1 \subset \ldots \subset M_r = M$ such that every quotient M_i/M_{i-1}

is isomorphic to the unit object $\mathcal{E}(1)$. The subcategory of unipotent difference modules over $\mathcal{P}$ is closed with respect to direct sums, duals, subquotients and tensor products. A unipotent module is called *decomposable* if it is a direct sum of two other modules. The following lemma gives the structure of the unipotent modules.

Lemma 6.1   1. *The cokernel of the map* $\phi - 1 : \mathcal{P} \to \mathcal{P}$ *has dimension 1. A basis of this space can be represented by* $t = z^{-1}$.

2. *For every* $n \geq 2$ *there exists a unique indecomposable unipotent module* M_n *of length* n. *The matrix of* Φ *with respect to a suitable basis of* M_n *has the form*

$$\begin{pmatrix} 1 & 0 & . & . & 0 \\ s & 1 & 0 & . & 0 \\ . & . & . & . & . \\ . & . & . & . & . \\ 0 & . & 0 & s & 1 \end{pmatrix}$$

where s *is any representative of the cokernel of* $\phi - 1$.

3. *The decomposition of a unipotent module* M *as a direct sum* $\oplus_{n \geq 1} M_n^{a_n}$ *is unique up to isomorphism. The integers* $a_n \geq 0$ *are unique and determine* M.

4. *Any unipotent difference equation has a fundamental matrix with coefficients in the simple difference ring* $\mathcal{P}[X]$, *given by* $\phi(X) = X + t$.

Proof: The proof of 1. is an easy calculation.

We prove 2. by induction on the length of M. A unipotent difference module of length 2 has a basis such that the matrix of Φ in this basis is $\begin{pmatrix} 1 & 0 \\ f & 1 \end{pmatrix}$. Let us call this module $M(f)$. Two elements $f_1, f_2 \in \mathcal{P}$ define isomorphic modules if and only if

$$f_1 = cf_2 + \phi(g) - g \text{ with } c \in C^* \text{ and } g \in \mathcal{P}$$

Letting s be a representative of the cokernel of $\phi - 1$, it follows that there are only two non isomorphic modules of length 2, namely $M(0)$ and $M(s)$. This proves the case $n = 2$ of part 2. of the lemma. We will prove the case $n = 3$ and omit the full proof of 2. By induction an indecomposable unipotent module of length 3 has a basis for which the matrix of Φ has the form $\begin{pmatrix} 1 & 0 & 0 \\ s & 1 & 0 \\ f & s & 1 \end{pmatrix}$ with $f \in$ $\mathcal{P}$. Let $B = \begin{pmatrix} 1 & 0 & 0 \\ c & 1 & 0 \\ g & 0 & 1 \end{pmatrix}$ with $g \in \mathcal{P}$ and $c \in C$.

Then $B^{-1} \begin{pmatrix} 1 & 0 & 0 \\ s & 1 & 0 \\ f & s & 1 \end{pmatrix} \phi(B)$ is a new matrix for Φ. A calculation shows that for suitable c and g this new matrix is the matrix of the lemma for $n = 3$. In order to verify that M_3 as given in the lemma is indecomposable one has to check that the equation $v = \begin{pmatrix} 1 & 0 & 0 \\ s & 1 & 0 \\ 0 & s & 1 \end{pmatrix} \phi(v)$ has a 1-dimensional solution space over C. This is an easy exercise.

Part 3. of the lemma follows from the Krull-Remak-Schmidt Theorem.

The first thing to prove for part 4. is two show that $\mathcal{P}[X]$ is simple. Let $f = X^d + a_{d-1} X^{d-1} + ... + a_0$ with $d > 0$ be such that $(f) \subset \mathcal{P}[X]$ is ϕ-invariant. Then $\phi(f) - f$ has degree less than d and must be 0. Comparing the coefficients of X^{d-1} of $\phi(f)$ and f one finds the equation $dt + \phi(a_{d-1}) = a_{d-1}$. This equation has no solution in $\mathcal{P}$. We conclude that $\mathcal{P}[X]$ is simple.

It suffices to show that M_n has a full set of solutions over $\mathcal{P}[X]$. This amounts to solving

$$\begin{pmatrix} v_1 \\ \cdot \\ \cdot \\ \cdot \\ v_n \end{pmatrix} = \begin{pmatrix} 1 & & & \\ t & 1 & & \\ \cdot & \cdot & \cdot & \\ \cdot & & \cdot & \cdot \\ & & t & 1 \end{pmatrix} \begin{pmatrix} \phi(v_1) \\ \cdot \\ \cdot \\ \cdot \\ \phi(v_n) \end{pmatrix}$$

or $\phi(v_1) - v_1 = 0$, $\phi(v_2) - v_2 = tv_1$, ..

One easily finds that v_1 is any constant, $v_2 = X v_1$, $v_3 = \frac{v_1}{2} X^2 +$ something in $\mathcal{P}$, etc. This proves part 4. ∎

There is a more efficient way to deal with unipotent difference modules. We will compare these modules with the differential modules over $\mathcal{P}$ which are nilpotent.

Let δ denote the differentiation on $\mathcal{P}$ given as $z \frac{d}{dz}$, i.e., $\delta(\sum a_\mu t^\mu) = \sum a_\mu (-\mu) t^\mu$. Let $\mathcal{P}[\delta]$ denotes the skew polynomial ring given by the formula $\delta a = a\delta + \delta(a)$. A differential module over $\mathcal{P}$ is a left $\mathcal{P}[\delta]$-module which is of finite dimension over $\mathcal{P}$. We write $\mathcal{E}$ for the unit object in the category of differential modules over $\mathcal{P}$. Thus $\mathcal{E} = \mathcal{P}e$ with $\delta e = 0$. A differential module M over $\mathcal{P}$ is called *nilpotent* if there exists a sequence of submodules $0 = M_0 \subset M_1 \subset ... \subset M_r = M$ such that every quotient M_i/M_{i-1} is isomorphic to the unit object $\mathcal{E}$. It is well known that every nilpotent differential module M has the form $M = \mathcal{P} \otimes_C W$, where W is a vector space over $\mathbf{C}$ invariant under the action of δ and such that the restriction of δ to W is a nilpotent linear map. We note that W is unique since it is the kernel of the operator δ^n on M for $n \geq$ the dimension of M over $\mathcal{P}$. In order to associate to a nilpotent differential module M a unipotent difference module, we have to introduce some notion of convergence.

The field $\mathcal{P}$ has a valuation given by $|f| = exp(-\lambda)$ if the expansion of $f \in \mathcal{P}^*$ starts with the term t^λ and of course $|0| = 0$. The field $\mathcal{P}$ is not complete with respect to this valuation but all the subfields $\mathbf{C}((t^{\frac{1}{m}}))$ are complete. A vector space N of finite dimension over $\mathcal{P}$ is given a norm by choosing a basis $e_1, ..., e_n$ and by defining $\| \sum a_i e_i \| = max \, |a_i|$. The topology induced by this norm does not depend on the chosen basis. Let M be a nilpotent differential module over $\mathcal{P}$. One defines an operator ϕ_M on M by the formula

$$\phi_M = exp \, (t\delta) = \sum_{n \geq 0} \frac{1}{n!} (t\delta)^n.$$

One can verify that $(t\delta)^{n+1} = t^{n+1}(\delta - n)(\delta - n + 1)...\delta$. This implies that the infinite expression for $\phi_M(m)$ converges for every $m \in M$. Further one can verify that the difference module M defined by $\Phi(m) = \phi_M(m)$ is a unipotent difference module.

On the other hand, a unipotent difference module M induces a nilpotent differential module by defining the action of δ on M by the formula

$$\delta = t^{-1} log \, (\Phi) := t^{-1} \sum_{n \geq 1} \frac{(-1)^{n+1}}{n} (\Phi - 1)^n.$$

The procedure above reflects of course the property that the automorphism ϕ of $\mathcal{P}$ is the exponential of the derivation $\frac{d}{dz} = -t^2 \frac{d}{dt}$ on $\mathcal{P}$.

The correspondence between unipotent difference modules and nilpotent differential modules is an equivalence of tensor categories as one easily sees.

As is well known nilpotent modules have a full set of solutions in the differential ring $\mathcal{P}[l]$, where $\frac{d}{dz} l := \frac{1}{z}$. In view of this correspondence with differential equations we prefer to consider the simple difference ring $\mathcal{P}[X]$ with $\phi(X) - X = log(1 + t)$. In the sequel we will write l for X and we note that l has as interpretation the multivalued function $log \, z$.

Theorem 6.2 *Every difference module M has a unique decomposition $\oplus_{g \in \mathcal{G}} M_g$, such that $M_g = \mathcal{E}(g) \otimes M(g)$ and $M(g)$ is a unipotent difference module. Moreover for $g_1 \neq g_2$ the vector space $Hom(M_{g_1}, M_{g_2})$ is 0.*

A proof of this statement follows easily from [21] and [47].

6.2 The universal Picard-Vessiot ring of $\mathcal{P}$

Define the ring $R = \mathcal{P}[\{e(g)\}_{g \in \mathcal{G}}][l]$ with generators $\{e(g)\}_{g \in \mathcal{G}}$ and l and relations $e(g_1 g_2) = e(g_1)e(g_2)$, $e(1) = 1$. Thus R is the polynomial ring in the variable

l over the group algebra of the group $\mathcal{G}$. The action of ϕ on R is given by $\phi(e(g)) = ge(g)$ and $\phi(l) - l = log(1 + t) \in \mathbb{C}\{t\} \subset \mathcal{P}$.

We will first show that R has only trivial ϕ-invariant ideals. Let I be a nonzero ideal of R with $\phi(I) \subset I$. Let $f \in I$ be a nonzero element such that its degree d as a polynomial in l is minimal. If $d > 0$, then one sees that $\phi(f) - f$ is a nonzero element with smaller degree. Hence $J := I \cap \mathcal{P}[e(g), g \in \mathcal{G}]$ is nonzero. Choose an element $f = \sum_{i=1,\ldots,r} a_i e(g_i) \neq 0$ in J with r minimal. If $r = 1$ then f is invertible and so I is the trivial ideal R. If $r > 1$ then after multiplying f with $a_1^{-1} e(g_1)^{-1}$ we may suppose that $a_1 = 1$ and $g_1 = 1$. Then $\phi(f) - f = \sum_{i=2,\ldots,r} (\phi(a_i)g_i - a_i)e(g_i) = 0$ by the definition of r. The term $a_i \in \mathcal{P}$ is not zero and satisfies $\phi(a_i)g_i = a_i$. This contradicts the construction of $\mathcal{G}$ and we have shown that R has only trivial ϕ-invariant ideals.

Next we will show that $\mathbb{C}$ is the ring of constants of R. Let f be a nonzero constant of R. The ideal (f) is ϕ-invariant and from the above it follows that f is a unit of R. In particular, f has degree 0 as polynomial in l and $f = \sum_{i=1,\ldots,r} a_i e(g_i)$ with $r \geq 1$ and all $a_i \neq 0$. The equation $\phi(f) = f$ and the definition of $\mathcal{G}$ easily implies that $f = ce(1)$ with $c \in \mathbb{C}^*$.

Every difference equation over $\mathcal{P}$ has a fundamental matrix with coefficients in R. Indeed, using the classification of Chapter 6.1, one finds that it suffices to consider the difference modules $\mathcal{E}(g)$ and the unipotent difference modules. For the first one $e(g)$ is the fundamental matrix. The unipotent case has been done in Lemma 6.1. Clearly R satisfies the minimality condition. This proves that R is indeed a universal Picard-Vessiot ring for $\mathcal{P}$.

We will now show that any universal Picard-Vessiot ring for $\mathcal{P}$ is isomorphic to the R above. Let B be another universal Picard-Vessiot ring and let $b(g)$ denote the invertible element of B satisfying $\phi(b(g)) = gb(g)$. Let L be an element of B satisfying $\phi(L) = L + log(1 + t)$. One can show that a normalization of the $b(g)$ exists such that $b(1) = 1$ and $b(g_1)b(g_2) = b(g_1 g_2)$ for all $g_1, g_2 \in \mathcal{G}$. Indeed, one considers the multiplicative subgroup H of B^* consisting of the elements of the form $ab(g)$, with $a \in \mathcal{P}^*$ and $g \in \mathcal{G}$. The surjective homomorphism $H \to \mathcal{G}$ has a right-inverse since $\mathcal{G}$ is a divisible group. Let $g \mapsto b(g)$ denote the right-inverse. Then clearly $b(g_1)b(g_2) = b(g_1 g_2)$. The obvious surjective, ϕ-equivariant $\mathcal{P}$-algebra homomorphism $F : R \to B$, $e(g) \mapsto b(g)$, $l \mapsto L$ has kernel 0 since R has only trivial ϕ-invariant ideals. This ends the proof.

The following result is now an easy consequence.

Corollary 6.3 *Let the difference field k be a difference subfield of $\mathcal{P}$ with field of constants C. The functor*

$$\omega : Diff(k, \phi) \to Vect_C$$

given by $\omega(M) = ker(\Phi - 1, R \otimes_k M)$ is an exact, faithful morphism of C-linear tensor categories.

As a final remark, we note that the ring R has zero divisors. This is due to the fact that the group $\mathcal{G}$ has elements of finite order. Indeed, the roots of unity μ_∞ in $C^* \subset \mathcal{G}$ is the torsion subgroup of $\mathcal{G}$. Put $e_m := e(e^{\frac{2\pi i}{m}})$. The factorization of $T^m - 1$ in linear polynomials yields zero divisors after substituting e_m for T.

6.3 Fields of constants which are not algebraically closed

Let $k \subset \mathcal{P}$ denote a difference subfield such that the field of constants k_0 of k is not algebraically closed. For convenience we will continue our discussion with the case $k = k_0(z)$ and C is the algebraic closure of k_0. There is a natural action of $Gal(C/k_0)$ on $\mathcal{P}$ given by the formula $\sigma(\sum a_\mu t^\mu) = \sum \sigma(a_\mu) t^\mu$. This action extends to the universal Picard-Vessiot ring R by posing $\sigma(l) = l$ and $\sigma(e(g)) = e(\sigma g)$. This action commutes with ϕ. For a difference module M over k the solution space $\omega(M) = ker(\Phi - 1, R \otimes_k M)$ is a C-vector space and is invariant under the action of $Gal(C/k_0)$. Using that $H^1(Gal(C/k_0), Gl(n, C))$ is trivial (See [58]) one finds that the k_0-vector space $\omega(M)^{Gal(C/k_0)}$ has the property $C \otimes_{k_0} \omega(M)^{Gal(C/k_0)} \to \omega(M)$ is an isomorphism. In other words $\omega(M)$ has a natural structure as $C \otimes_{k_0} \tilde{\omega}(M)$ where $\tilde{\omega}(M)$ is a vector space over k_0. We will call this a k_0-structure on $\omega(M)$. Now $M \to \tilde{\omega}(M)$ defines a functor from the category of difference modules over k to the category of vector spaces (of finite dimension) over k_0. As in Corollary 6.3 this is an exact, faithful morphism of k_0-linear tensor categories. In particular, the Tannakian approach yields a difference Galois group for M, defined over k_0, such that $G \otimes_{k_0} C$ is the difference Galois group of $C(z) \otimes M$. The k_0-structure on $\omega(M)$ has also consequences for rationality properties and algorithms concerning difference equations. (Compare [28]).

6.4 Automorphisms of the universal Picard-Vessiot ring of $\mathcal{P}$

For a homomorphism $h : \mathcal{G} \to C^*$ and a constant $c \in C$ one defines the automorphism σ of R over $\mathcal{P}$, commuting with ϕ, by $\sigma(e(g)) = h(g)e(g)$ and $\sigma(l) = l + c$. This describes the group of all such automorphisms. The group is commutative. As group scheme over C it has the coordinate ring $C[e(g)|g \in \mathcal{G}][l]$ with co-multiplication m given by $m(e(g)) = e(g) \otimes e(g)$ and $m(l) = l \otimes 1 + 1 \otimes l$.

6.5   Difference equations over $C((z^{-1}))$ and the formal Galois group.

The Galois group of a difference equation over the field $C((z^{-1}))$ will be called the *formal Galois group*. In order to find this group we want to find first the universal Picard-Vessiot ring Ω of $C((z^{-1}))$.

The ring Ω lies in R, the universal Picard-Vessiot ring of $\mathcal{P}$. It is clear that Ω must contain $\mathcal{P}$ and l. Any $g \in \mathcal{G}$ lies in some finite extension of $C((z^{-1}))$ and has finitely many conjugates. One can then construct a difference equation over $C((z^{-1}))$ with eigenvalues the conjugates of g. This shows that Ω must coincide with R. The group of automorphisms of R over $C((z^{-1}))$, commuting with ϕ, is denoted by $Aut(R/C((z^{-1})), \phi)$. One finds a split exact sequence of groups

$$1 \to Aut(R/\mathcal{P}, \phi) \to Aut(R/C((z^{-1})), \phi) \to Gal(\mathcal{P}/C((z^{-1}))) \to 1$$

Any automorphism σ of $\mathcal{P}$ over $C((z^{-1}))$ acts on $\mathcal{G}$. The natural extension of σ to an element of $Aut(R/C((z^{-1})), \phi)$ is given by $\sigma(l) = l$ and $\sigma(e(g)) = e(\sigma g)$. This formula defines the splitting. The group $Aut(R/C((z^{-1})), \phi)$ is not commutative.

We will describe the Galois group of a difference module (or equation) M over $C((z^{-1}))$ in case $C = \mathbf{C}$ is the field of complex numbers. The solution space of this difference module is $V := ker(\Phi - 1, R \otimes_{\mathbf{C}((z^{-1}))} M)$. The automorphism group $Aut(R/\mathbf{C}((z^{-1})), \phi)$ acts on V and the image of this group is the formal Galois group of M.

One can make this more explicit by defining a special element $\gamma \in Aut(R/\mathbf{C}((z^{-1})), \phi)$, called *the formal monodromy*. This γ acts on $\mathcal{P}/\mathbf{C}((z^{-1}))$ by $\gamma(z^\lambda) = e^{2\pi i \lambda} z^\lambda$ for all $\lambda \in \mathbf{Q}$. The action of γ on $\mathcal{G}$ is induced by its action on $\mathcal{P}^*$. Finally, γ acts on R by $\gamma e(g) = e(\gamma g)$ and $\gamma(l) = l + 2\pi i$. The image of γ in $Gl(V)$ will be denoted by γ_V and will be called the *formal monodromy* of the differential module M.

The group $Aut(R/\mathcal{P}[l], \phi)$ and its image $\mathcal{T}$ in $Gl(V)$ will be called the *exponential torus*.

The group generated by γ and $Aut(R/\mathcal{P}[l], \phi)$ is Zariski-dense in $Aut(R/\mathbf{C}((z^{-1})), \phi)$. This has as consequence that the Zariski closure of the subgroup of $Gl(V)$, generated by γ_V and $\mathcal{T}$ is equal to the formal Galois group of M. This is completely analogous to the differential case: "The formal differential Galois group of a differential module is generated as algebraic group by the exponential torus and the formal monodromy." (See [40]). One can use the method of [51] to translate the classification of difference modules over $\mathbf{C}((z^{-1}))$ and its formal Galois group in terms of linear algebra. This works as follows:

One considers a category $\mathcal{F}$ with objects of the form $(V, \oplus V_g, \gamma_V)$, where:
(a) V is a finite dimensional vector space over $\mathbf{C}$.

(b) V has a direct sum decomposition $V = \oplus_{g \in \mathcal{G}} V_g$.

(c) $\gamma_V \in GL(V)$ is supposed to satisfy:

1. $\gamma_V(V_g) = V_{\gamma g}$, where $\gamma \in Aut(R/\mathbf{C}((z^{-1})), \phi)$ is the formal monodromy.

2. Let $d \geq 1$ be the smallest integer such that $\gamma^d g = g$ holds for all $g \in \mathcal{G}$ with $V_g \neq 0$. Then γ_V^d is unipotent.

A morphism $A : (V, \oplus V_g, \gamma_V) \to (W, \oplus W_g, \gamma_W)$ is a $\mathbf{C}$-linear map $A : V \to W$ such that $A(V_g) \subset W_g$ for all $g \in \mathcal{G}$ and $\gamma_W A = A \gamma_V$. The category $\mathcal{F}$ has a natural structure as Tannakian category. The forgetful functor $\omega : (V, \oplus V_g, \gamma_V) \mapsto V$ from $\mathcal{F}$ to $Vect_{\mathbf{C}}$ is a fibre functor. Thus $\mathcal{F}$ is a neutral Tannakian category. For a fixed object $(V, \oplus V_g, \gamma_V)$ of $\mathcal{F}$, one considers the full Tannakian subcategory generated by this object. The Tannakian group of the object $(V, V_{g_1} \oplus \ldots \oplus V_{g_s}, \gamma_V)$, i.e., the group scheme corresponding to this subcategory, is easily seen to be the algebraic subgroup of $Gl(V)$ generated (as algebraic group) by γ_V and by the group of the linear maps $\oplus_{i=1}^{s} \lambda_i id_{V_{g_i}}$, with $\lambda_1, ..., \lambda_s \in \mathbf{C}^*$ satisfying: if $g_1^{n_1}...g_s^{n_s} = 1$ in $\mathcal{G}$ then $\lambda_1^{n_1}...\lambda_s^{n_s} = 1 \in \mathbf{C}^*$.

To a difference module M over $\mathbf{C}((z))$ we associate $(V, \oplus V_g, \gamma_V)$ defined by $V = \omega(M) = ker(\Phi - 1, R \otimes M)$; $V_g = V \cap (\mathcal{P}[l]e(g) \otimes M)$ and γ_V is the action of γ to V. It can be seen that this induces an equivalence from the category of the difference modules over $\mathbf{C}((z^{-1}))$ to the category $\mathcal{F}$. This equivalence preserves tensor products and is an equivalence of Tannakian categories. In particular, the formal Galois group of a difference module M coincides with the Tannakian group of the associated $(V, \oplus V_g, \gamma_V)$. In other words, the formal Galois group is, as algebraic group, generated by the formal monodromy γ_V and the group $\oplus_{i=1}^{s} \lambda_i id_{V_{g_i}}$, with $\lambda_1, ..., \lambda_s \in \mathbf{C}^*$ satisfying: if $g_1^{n_1}...g_s^{n_s} = 1$ in $\mathcal{G}$ then $\lambda_1^{n_1}...\lambda_s^{n_s} = 1 \in \mathbf{C}^*$.

The last group is the exponential torus, as defined above.

Strictly speaking $\mathcal{T}$ is not always an algebraic torus since $\mathcal{T}$ need not be connected. The Galois group G of M is easily seen to have the following properties:

1. G/G^o is commutative and generated by at most two elements.

2. There exists a $B \in Gl(V)$ such that G^o is the Zariski-closure of the group generated by B.

We note that the term *formal monodromy* above is appropriate. Indeed, M has a fundamental matrix Y with coefficients in R. The elements $e(g)$ and l of R have an interpretation as multivalued functions. Hence Y can be interpreted as a matrix whose coefficients are combinations of functions in $\mathbf{C}((z^{-1}))$ and $\Gamma^\lambda, c^z, z^{a_0}, exp(\sum_{0 < \mu < 1} a_\mu z^\mu), log\ z$. Analytic continuation on a circle around ∞ coincides with the formal monodromy.

Analytic Theory

The analytic aspects of ordinary difference equations over the fields $\mathbf{C}(z)$ and $\mathbf{C}(\{z^{-1}\})$ arise from the problem of interpreting a symbolic solution (or a symbolic fundamental matrix) as an actual solution involving meromorphic functions defined on certain regions in the complex plane. For certain difference equations, namely the "semi-regular" equations, the asymptotic theory is elementary. There are unique asymptotic lifts F_{right} and F_{left} of the symbolic fundamental matrix. Those lifts are defined above a "right domain" and a "left domain" and involve meromorphic functions and the symbols $\{e(c)|c \in \mathbf{C}^*,\ c$ a root of unity$\}$. The comparison of the two lifts leads to two connection matrices, defined on an upper and on a lower half plane. The two connection matrices are given by the same formula $F_{right}^{-1}F_{left}$. Their coefficients lie in the algebraic closures of the fields $\mathbf{C}(\{u\})$ and $\mathbf{C}(\{u^{-1}\})$, where u stands for $e^{2\pi i z}$. The connection matrices are canonical in the sense that they are functorial and commute with tensor products. The difference Galois group can be expressed in terms of the formal difference Galois group and the connection matrices. Using this one shows that any linear algebraic group G with G/G^o cyclic is the difference Galois group of a semi-regular equation over the field $k_\infty = \mathbf{C}(\{z^{-1}\})$, the field of convergent Laurent series at ∞. We also discuss the problem of constructing equations over $\mathbf{C}(z)$ with prescribed difference Galois group.

For a more general type of difference equations, namely the "mild" equations, the regions where one can find asymptotic lifts of a symbolic fundamental matrix are more restricted. Moreover the lifts are in general not unique. The method of multisummation for difference equations, introduced in work of B.L.J. Braaksma and B.F. Faber, produces however unique lifts of a special nature. Using that multisummation is a functorial process and commutes with tensor products, one can again define canonical connection matrices. The difference Galois group is again expressed in terms of the formal difference Galois group and the connection matrices. This leads to a conjectural description of the linear algebraic groups which are the difference Galois group of a difference equation over k_∞. We partially verify this conjecture.

Another interesting aspect is the strong and intricate relation between "mild" differential equations and mild difference equations over the field $\mathbf{C}(\{z^{-1}\})$. This leads to a relation between the connection matrices of the difference equation and the Stokes matrices of the differential equation.

For general "wild" difference equations the asymptotic theory is more involved. We give an exposition of this theory based on work of B.L.J. Braaksma, B.F. Faber and G.K. Immink. The aim is to find large sectors for which asymptotic lifts exist. The theorem of Malgrange and Sibuya is also the main tool for the classification of the difference equations over k_∞ with prescribed formal structure.

In the last section, these analytic aspects are also pursued for the case of q-difference equations over the fields $\mathbf{C}(z)$ and $\mathbf{C}(\{z^{-1}\})$.

It was a surprise to discover in the work of G.D. Birkhoff and W.J. Tritzinsky much of what is presented here. In particular, Birkhoff's paper [9] of 1911 essentially presents the connection matrix for regular singular equations. His definition of the connection matrix (or matrices) is not the one given here. This difference is due to their interpretation of the symbols $e(e^{2\pi i r})$, with $r \in \mathbf{Q}$ as $e^{2\pi i r z}$ and to different choices of $\log(z)$. Part of the work presented here can be seen as a modern version of [9]. The relation between mild differential equations and mild difference equation is the subject of Birkhoff's paper [12]. We reprove part of this paper and point out a mistake. The paper [14] deals with what we call "wild" difference equations. The "Fundamental Existence Theorem" gives the lifting of symbolic solutions of a difference equation to "quadrants". The work of G.K. Immink and our presentation of that is again a modern version of [14].

Chapter 7

Classification and canonical forms

7.1   A classification of singularities

The singularity at infinity of a difference equation $y(z + 1) = A\, y(z)$ over k, where k is one of the fields $\mathbf{C}(z)$, $k_\infty := \mathbf{C}(\{z^{-1}\})$ or $\hat{k}_\infty := \mathbf{C}((z^{-1}))$, will be classified in a rough way by the terms: *regular, regular singular, very mild, mild* and *wild*. The terminology reflects the asymptotic theory of the equation. The classification can be read of from the formal (i.e. over $\mathbf{C}((z^{-1}))$) classification of the equation. We recall that this formal classification involves the subgroup $\mathcal{G}$ of $\mathcal{P}^*$, where $\mathcal{P}$ is the algebraic closure of $\hat{k}_\infty$, consisting of the elements

$$z^\lambda c\; exp(\phi(q) - q)(1 + z^{-1})^{a_0}$$

with $\lambda \in \mathbf{Q}$, $c \in \mathbf{C}^*$, $a_0 \in L \subset \mathbf{C}$ and q is a finite sum $\sum_{0 < \mu < 1} a_\mu z^\mu$ with $a_\mu \in \mathbf{C}$ and $\mu \in \mathbf{Q}$. The term L denotes a $\mathbf{Q}$-linear vector space of $\mathbf{C}$ such that $L \oplus \mathbf{Q} = \mathbf{C}$.

The elements of $\mathcal{G}$ occuring in the formal decomposition of the equation (over the algebraic closure $\mathcal{P}$ of $\hat{k}_\infty$) are called the *eigenvalues* of the equation (see Theorem 6.2). The equation is called *wild* if an eigenvalue with $\lambda \neq 0$ is present and *mild* if this is not the case. The equation is *very mild* if for every eigenvalue one has $\lambda = 0$ and $c = 1$. If the only eigenvalues are of the form $(1 + z^{-1})^{a_0}$ then the equation is called *regular singular*. Finally the equation is called *regular* if the equation is trivial over $\hat{k}_\infty$.

Lemma 7.1 *The following properties of $y(z+1) = A\, y(z)$ over k are equivalent.*

1. *The equation is regular.*

2. *There is a formal fundamental matrix* $F = F(z) \in Gl(n, \hat{k}_\infty)$, *i.e* $F(z + 1) = AF(z)$.

3. *There is an equivalent equation* $v(z + 1) = \tilde{A}\, v(z)$ *with* $\tilde{A} := B(z + 1)^{-1}AB(z)$ *and* $B \in Gl(n, k)$ *such that the expansion of* $\tilde{A}$ *at infinity reads* $1 + \tilde{A}_2 z^{-2} + \tilde{A}_3 z^{-3} + \ldots$.

Proof: $(1){\Leftrightarrow}(2)$ is obvious. Suppose that there is a formal fundamental matrix F. One can approximate F by some $B \in Gl(n, k)$ such that $\tilde{A} := B(z + 1)^{-1}AB(z)$ has the required form. This shows $(2){\Rightarrow}(3)$. If A has the expansion $1 + A_2 z^{-2} + A_3 z^{-3} + \ldots$ at infinity then one easily shows that there is a unique formal matrix F of the form $1 + F_1 z^{-1} + F_2 z^{-2} + \ldots$ such that $F(z+1) = AF(z)$. This proves $(3){\Rightarrow}(2)$. ∎

The difference ring $\hat{k}_\infty[\{Z^a\}_{a \in \mathbf{C}}, l]$ is defined by the relations $Z^a Z^b = Z^{a+b}$, $Z^1 = z \in \hat{k}_\infty$ and the action of ϕ with $\phi(Z^a) = (1 + z^{-1})^a Z^a$; $\phi(l) = l + \log(1 + z^{-1})$. This ring is in fact a subring of the universal Picard-Vessiot ring defined in Section 6.2.

Lemma 7.2 *The following properties of* $y(z+1) = A\, y(z)$ *over* k *are equivalent.*

1. *The equation is regular singular.*

2. *The equation is trivial over the ring* $\hat{k}_\infty[\{Z^a\}_{a \in \mathbf{C}}, l]$.

3. *There is a fundamental matrix with coefficients in* $\hat{k}_\infty[\{Z^a\}_{a \in \mathbf{C}}, l]$.

4. *There is a formal* $F \in Gl(n, \hat{k}_\infty)$ *and a constant matrix* D *such that* $F(z + 1)^{-1}AF(z) = (1 + z^{-1})^D$.

5. *There is an equivalent equation* $v(z + 1) = \tilde{A}\, v(z)$ *with* $\tilde{A} := B(z + 1)^{-1}AB(z)$ *and* $B \in Gl(n, k)$ *such that the expansion of* $\tilde{A}$ *at infinity is* $1 + \tilde{A}_1 z^{-1} + \tilde{A}_2 z^{-2} + \ldots$.

Proof: Most of the implications follow from the formal classification of difference modules. We will show here (as an example) that (5) implies (4).

We may suppose that A has already the expansion $1 + A_1 z^{-1} + \ldots$ at ∞. The matrix A_1 is supposed to be in Jordan normal form. Let a Jordan block with eigenvalue λ be given. The matrix B is supposed to be a diagonal matrix with z's on the diagonal of this Jordan block and 1's on the diagonal of the other blocks. Let $\tilde{A} = B(z + 1)^{-1}AB(z) = 1 + \tilde{A}_1 z^{-1} + \ldots$. Then $\tilde{A}_1$ is almost the same as A_1. The only change is that λ is replaced by $\lambda - 1$. With operations of this type one can transform the equation to a new equation, again called $y(z + 1) = A\, y(z)$, such that every eigenvalue λ of A_1 satisfies $0 \le Re(\lambda) < 1$. In particular, the differences of distinct eigenvalues of A_1 are not integral.

This last property is used in the proof that there exists a unique formal $F = 1 + F_1 z^{-1} + F_2 z^{-2} + \ldots$ such that $F(z + 1)^{-1} A F(z) = (1 + z^{-1})^D$ with $D = A_1$. ∎

Recall that the algebraic closure of $\hat{k}_\infty$ is denoted by $\mathcal{P}$. This is the field of formal Puiseux series over $\mathbf{C}$ in the variable z^{-1}. The following two lemmas are also easy consequences from the formal classification of difference equations.

Lemma 7.3 *The following conditions are equivalent*

1. *The equation $y(z + 1) = A\, y(z)$ is very mild.*

2. *The equation is equivalent (over k) with an equation $v(z+1) = \tilde{A}\, v(z)$ with expansion $\tilde{A} = \tilde{A}_0 + \tilde{A}_1 z^{-1} + \ldots$ at infinity such that $\tilde{A}_0 - 1$ is nilpotent.*

Lemma 7.4 *The following conditions are equivalent*

1. *The equation $y(z + 1) = A\, y(z)$ is mild.*

2. *The equation is equivalent (over k) with an equation $v(z + 1) = \tilde{A}\, v(z)$ with expansion $\tilde{A} = \tilde{A}_0 + \tilde{A}_1 z^{-1} + \ldots$ at infinity such that $\tilde{A}_0$ is invertible.*

3. *There is a fundamental matrix with coefficients in $\mathcal{P}[\{e(g)\}_{g \in \mathcal{G}_{mild}}, l]$, where $\mathcal{G}_{mild}$ denotes the subgroup of $\mathcal{G}$ consisting of the elements*

$$\{c\, exp(\phi(q) - q)(1 + z^{-1})^{a_0} | c \in \mathbf{C}^*,\ a_0 \in L,\ q = \sum_{0 < \mu < 1} a_\mu z^\mu\}.$$

Over the field $\hat{k}_\infty$ one can compare differential modules and difference modules. A differential module M is a vector space over $\hat{k}_\infty$ of finite dimension, together with a $\mathbf{C}$-linear action of $\frac{d}{dz}$ on M, satisfying the formula $\frac{d}{dz}(fm) = f\frac{d}{dz}(m) + (\frac{df}{dz})m$ for $f \in \hat{k}_\infty$ and $m \in M$. The well known formal classification of differential modules over the algebraic closure $\mathcal{P}$ of $\hat{k}_\infty$ can be formulated as follows:

One considers $\mathcal{Q}$, the group (or vector space over $\mathbf{Q}$), defined as

$$\{\sum_{\mu \geq 0} a_\mu z^\mu | \text{ finite sums },\ \mu \in \mathbf{Q},\ a_\mu \in \mathbf{C}\}/\mathbf{Q}.$$

Let L denote a $\mathbf{Q}$-vector space with $L \oplus \mathbf{Q} = \mathbf{C}$. Then we may identify $\mathcal{Q}$ with the space

$$\{\sum_{\mu \geq 0} a_\mu z^\mu | \text{ finite sums },\ \mu \in \mathbf{Q},\ a_\mu \in \mathbf{C},\ a_0 \in L\}.$$

For $q \in \mathcal{Q}$, one defines the one dimensional differential module $\mathcal{D}(q) = \mathcal{P}e_q$ over $\mathcal{P}$ by $\frac{d}{dz}e_q = z^{-1}qe_q$. For a differential module M over $\mathcal{P}$ there is a unique decomposition $M = \oplus_{q\in\mathcal{Q}}\mathcal{D}(q) \otimes M(q)$ where the $M(q)$ are nilpotent differential modules over $\mathcal{P}$. The set $\{q \in \mathcal{Q}|\ M(q) \neq 0\}$ is called the set of *eigenvalues* of M. A differential module M over $\hat{k}_{\infty}$ is called *mild* if the eigenvalues of $\mathcal{P} \otimes M$ are contained in

$$\{ \sum_{0\leq\mu\leq 1} a_{\mu}z^{\mu}|\ \text{finite sums} \ , \mu \in \mathbf{Q},\ a_{\mu} \in \mathbf{C},\ a_0 \in L\}.$$

M is called *very mild* is this set of eigenvalues is contained in

$$\{ \sum_{0\leq\mu< 1} a_{\mu}z^{\mu}|\ \text{finite sums} \ , \mu \in \mathbf{Q},\ a_{\mu} \in \mathbf{C},\ a_0 \in L\}.$$

A lattice $M_0 \subset M$ is a free module over the ring of integers $\mathbf{C}[[z^{-1}]]$ of $\hat{k}_{\infty}$ such that the natural map $\hat{k}_{\infty} \otimes_{\mathbf{C}[[z^{-1}]]} \hat{k}_{\infty} \to M$ is an isomorphism. One can show that M is mild if and only if there is a lattice M_0 with $\frac{d}{dz}M_0 \subset M_0$. The translation into matrix differential equations is: M is mild if and only M corresponds to an equation $\frac{d}{dz}y = A\ y$ where A has expansion $A_0 + A_1z^{-1} + \ldots$ at infinity. Furthermore "M very mild" is equivalent with the existence of a lattice M_0, such that $\frac{d}{dz}M_0 \subset M_0$ and $\frac{d}{dz}$ acts as a nilpotent map on $M_0/z^{-1}M_0$. The translation into matrix differential equations is: M is very mild if and only M corresponds to an equation $\frac{d}{dz}y = A\ y$ where A has expansion $A_0 + A_1z^{-1} + \ldots$ at infinity with A_0 nilpotent.

On a mild differential module M one defines the operator Φ as the infinite sum $exp(\frac{d}{dz}) = \sum \frac{1}{n!}(\frac{d}{dz})^n$. It can be seen, by using the topology of $\mathbf{C}$, that this infinite expression makes sense and defines a difference module. This difference module is mild. Using the formal classification of difference modules one easily sees that every mild difference module is obtained in this way from a mild differential module. However this mild differential module is far from unique.

The situation is more transparent for very mild difference modules. For such a module one easily shows that

$$\frac{d}{dz} := \log(\Phi) := \sum_{n\geq 1} \frac{(-1)^{n+1}}{n}(\Phi - 1)^n$$

converges on M and defines on M the structure of a very mild differential module.

The construction above gives an equivalence between the abelian tensor categories of very mild difference modules and very mild differential modules. The equivalence maps regular (regular singular resp.) difference modules to regular (regular singular resp.) differential modules.

7.2 Canonical forms

A difference equation $y(z + 1) = A\ y(z)$ can formally be transformed into a canonical form $A^c = F(z + 1)^{-1} A F(z)$. Usually F is taken to be an invertible matrix with coefficients in $\mathcal{P}$ and the canonical form A^c is a direct sum of blocks $g(1 + M_g \log(1 + z^{-1}))$ where g runs over a finite subset of $\mathcal{G}$ and with each M_g a constant nilpotent matrix. For our purposes this is convenient and sufficient. However the finite extension of k_∞ needed for F and A^c might cause problems. In this section we will derive a canonical form and a transformation which are defined over k_∞ and $\hat{k}_\infty$. We fix an element $g_0 \in \mathcal{G}$. In general the one dimensional module with generator $e(g_0)$ and relation $\phi(e(g_0)) = g_0 e(g_0)$ is not defined over k_∞ since g_0 lies in a finite extension of k_∞. Let T be the smallest field extension of k_∞ such that $g_0 \in T$. Then $Te(g_0)$ is a well defined difference module over T. Let g_i, $i = 0, 1, \ldots, m - 1$ denote the conjugates of g_0 in T. By minimality $m = [T : k_\infty]$. Consider now the difference module $M := Te(g_0) + \ldots + Te(g_{m-1})$. The generator of the Galois group of T/k_∞ will be denoted by γ. We define γ, as usual, by $\gamma(z^{1/m}) = e^{2\pi i/m} z^{1/m}$. We may suppose that $\gamma^j(g_0) = g_j$. We let γ act on M by the prescribed action on T and $\gamma(e(g_i)) = e(g_{i+1})$. For convenience we have introduced the notation $g_{i+km} = g_i$ for $0 \leq i < m - 1$ and any k. The action of γ commutes with the operation of Φ on M. Let $E(\tilde{g}_0)$ denote the set of γ-invariant elements of M. This is a vector space over k_∞ and also a difference module. An element $\sum_{j=0}^{m-1} a_j e(g_j)$, with $a_j \in T$, is invariant under γ if and only if $a_{j+1} = \gamma a_j$ for all j. Put $\zeta = e^{2\pi i/m}$. An explicit basis of $E(\tilde{g}_0)$ over k_∞ is

$$z^{k/m} \sum_{j=0}^{m-1} \zeta^{kj} e(g_j) \text{ for } k = 0, \ldots, m - 1.$$

This shows that $T \otimes E(\tilde{g}_0) \to M$ is an isomorphism. The matrix of Φ on this basis has coefficients in k_∞. We use the notation $\tilde{g}$ for $g \in \mathcal{G}$ to denote an element of the orbit of g under the action of the Galois group of the algebraic closure of k_∞ over k_∞. A *canonical difference module over k_∞* if defined to be a direct sum (over the orbits in $\mathcal{G}$), $\oplus E(\tilde{g}) \otimes M_{\tilde{g}}$ where each $M_{\tilde{g}}$ is a unipotent difference module (i.e. with a matrix of the sort $1 + z^{-1} N$ for some nilpotent matrix N). It is clear that every difference module over k_∞ is formally, i.e. over $\hat{k}_\infty$, equivalent with a *unique* canonical difference module over k_∞. One can translate this in terms of matrices. Every canonical difference module over k_∞ corresponds to an invertible matrix A^c with coefficients in k_∞. This matrix can be computed by using the given basis of $E(\tilde{g})$ and by choosing a standard form for nilpotent matrices. For every difference equation $y(z + 1) = A\ y(z)$ there is a unique canonical matrix A^c and a unique invertible matrix F with coefficients in $\hat{k}_\infty$ such that $F(z + 1)^{-1} A F(z) = A^c$.

One might hope that for difference equations over $\mathbf{C}(z)$ there is also a "canonical form" defined over $\mathbf{C}(z)$ which is over k_∞ equivalent with the original one. However the canonical difference modules $E(\tilde{g})$ considered above are, in general, not equivalent over k_∞ with modules defined over $\mathbf{C}(z)$. (See examples 9.8 and 9.9). A possibility would be to replace every $E(\tilde{g})$ by some module $E'(\tilde{g})$ which is now defined over $\mathbf{C}(z)$ and is over $\hat{k}_\infty$ equivalent with $E(\tilde{g})$. A truncation of $E(\tilde{g})$ has this property. There seems to be no evident choice for the $E'(\tilde{g})$. Unlike the case of differential modules, truncations of $E'(\tilde{g})$ with respect to high orders of z^{-1} are not isomorphic over k_∞ (Over $\hat{k}_\infty$ they are isomorphic). In particular, the asymptotic theories, corresponding to different choices of truncations, are different.

Chapter 8

Semi-regular difference equations

8.1 Introduction

The asymptotic theory of general difference equations has some analogies with the theory for differential equations. However, difference equations are more difficult to handle. In later sections we will study the asymptotic theory, connection matrices and the relations with the difference Galois groups for general difference equations. As a guide to this we will work out this theory for special equations, the semi-regular difference equations. For those equations one can construct asymptotic lifts "by hand". Moreover those lifts are unique and are defined on large sectors. The relation with difference Galois groups is rather transparent for semi-regular equations.

A difference equation over k_∞ (or over $\mathbf{C}(z)$ or $\hat{k}_\infty$) is called *semi-regular of type n* if the equation is formally equivalent (i.e. over $\hat{k}_\infty$) to a matrix equation $y(z + 1) = A\, y(z)$ with a diagonal matrix A, having nth roots of unity on the diagonal.

The following properties are equivalent:

(1) The equation is semi-regular of type n.

(2) The "eigenvalues" of the difference equation are contained in $\mu_n :=$ the group of the nth roots of unity in $\mathbf{C}^*$.

(3) The equation has a full set of solutions in the difference ring $\hat{k}_\infty[e(c)]_{c\in\mu_n}$. We will write this ring as $k_\infty[\mu_n]$.

The notion of "semi-regular of type 1" coincides with "regular". A difference module is called semi-regular if it is semi-regular of type n for some $n \geq 1$. The semi-regular (of type n) difference modules over k_∞ form a full subcategory of the category of all difference modules over k_∞. The tensor product of two semi-regular modules (of type n) is again semi-regular (of type n). Further quotient modules and submodules of a semi-regular module (of type n) are semi-regular (of type n).

8.2   Some easy asymptotics

We recall the definition of *asymptotic expansion*. Let a sector at ∞ be defined by $\{z \in \mathbf{C} |\ |z| > R,\ \arg(z) \in (a, b)\}$. A holomorphic function f on this sector has the asymptotic expansion $\sum_{n \geq A} a_n z^{-n} \in \hat{k}_\infty = \mathbf{C}((z^{-1}))$ if there exists for every $N > A$ and every positive ϵ a constant $C > 0$ such that

$$|f(z) - \sum_{A \leq n < N} a_n z^{-n}| \leq C |z|^{-N}$$

holds for all z with $|z| \geq R + \epsilon$ and $\arg(z) \in (a + \epsilon, b - \epsilon)$.

The first lemma that we consider is rather elementary. This lemma suffices for the construction of the connection matrix of a semi-regular equation. We consider ordinary difference equations in the neighborhood of ∞. As before, the field of convergent Laurent series at infinite is denoted by $k_\infty = \mathbf{C}(\{z^{-1}\})$ and the field of formal Laurent series in z^{-1} is denoted by $\hat{k}_\infty$. A subset of the complex plane of the form

$$\{z \in \mathbf{C} |\ |Im(z)| > R \text{ or } (|z| > R \text{ and } Re(z) > 0)\}$$

will be called a *right domain*. A subset U of $\mathbf{C}$ is called a *left domain* if $-U$ is a right domain.

Lemma 8.1 *Let $\dot{y}$ be a formal solution (i.e. with coefficients in $\hat{k}_\infty$) of the equation $y(z + 1) = A\ y(z) + a$, where A and a have coefficients in k_∞. The expansion of A at ∞ is supposed to have the form $A_0(1 + A_1 z^{-1} + A_2 z^{-2} + \ldots)$ where A_0 is a semi-simple matrix such that all its eigenvalues have absolute value 1. There exists a unique meromorphic vector y_{right}, such that:*

(1) y_{right} is defined on a right-domain.

(2) y_{right} has $\dot{y}$ as asymptotic expansion at ∞ in a sector with $\arg(z) \in (-\pi + \epsilon, +\pi - \epsilon)$ for every $\epsilon > 0$.

(3) $y_{right}(z + 1) = A\ y_{right}(z) + a$.

Proof: The norms that we will use for matrices M and vectors v are the usual ones, namely $\|M\| = \sqrt{\sum |M_{i,j}|^2}$ and $\|v\| = \sqrt{\sum |v_i|^2}$. We have chosen coordinates such that A_0 is a diagonal matrix. The norm of A_0 is therefore 1. Choose some meromorphic vector y_1 which has asymptotic expansion $\hat{y}$ on any sector at ∞ with opening $(-\pi + \epsilon, \pi - \epsilon)$ for every $\epsilon > 0$. Put $y_2 = y_1 + g$, where g should be chosen such that g has asymptotic expansion 0 at the same sectors at ∞ and such that $y_2(z + 1) = A\, y_2(z) + a$. This means that g should be a solution of

$$g(z + 1) - A(z)g(z) = -(y_1(z + 1) - A(z)y_1(z) - a) := -b(z),$$

where (by construction) $b(z)$ is a meromorphic vector which has asymptotic expansion 0 in the same sectors at ∞. One proposes the solution

$$g(z) = \sum_{n=0}^{\infty} A(z)^{-1}A(z + 1)^{-1}\ldots A(z + n)^{-1}b(z + n).$$

Consider a bounded open sector $S \subset \mathbf{C}$ with $arg(z) \in (-\pi + \epsilon, \pi - \epsilon)$. On S one has the following inequalities:

(1) There is a constant $c > 0$ such that for all $z \in S$ one has

$$c \max(|z|, n) \le |z + n| \le 2\max(|z|, n)$$

(2) $\|A_0 A(z)^{-1}\| \le (1 + d|z|^{-1})$ for some constant $d > 0$.

(3) $\|b(z)\| \le \frac{C_s}{|z|^s}$ for all $s = 1, 2, \ldots$ and some constants C_s.

Take $s \gg 0$. We then have

$$\|A(z)^{-1}\ldots A(z + n)^{-1}b(z + n)\| \le (1 + \frac{D}{|z|})\ldots(1 + \frac{D}{|z| + n}) \frac{C_s}{|z + n|^s},$$

where $D = [2dc^{-1}] + 1$. The estimate in this formula is equal to

$$(1 + \frac{n + 1}{|z|})\ldots(1 + \frac{n + 1}{|z| + D - 1}) \frac{C_s}{|z + n|^s}.$$

We may assume that $|z| \ge 1$ holds on the sector S. Using this and the inequality (1) one finds an estimate

$$\|A(z)^{-1}\ldots A(z + n)^{-1}b(z + n)\| \le C_s c^{-s} \frac{(n + 1)^D}{n^{D+2}} \frac{1}{|z|^{s-D-2}}.$$

The infinite sum therefore converges uniformly on S. Further one finds the estimate

$$|g(z)| \le C_s c^{-s} 2^D \sum_{n=1}^{\infty} \frac{1}{n^2} \frac{1}{|z|^{s-D-2}}.$$

This shows that g has asymptotic expansion 0 in S and therefore also on the sectors with $\arg(z) \in (-\pi + \epsilon, \pi - \epsilon)$. Finally, by construction g satisfies the required equation.

The vector y_2 exists on some sector at ∞ with opening $(-\pi, \pi)$ and satisfies conclusions (2) and (3). The equation 3. shows that y_2 has an analytic continuation y_{right} on a right-domain.

Suppose that y'_{right} also satisfies conclusions (1), (2) and (3). The difference v of y_{right} and y'_{right} has asymptotic expansion 0. The formula $v(z+1) = A\, v(z)$ implies $v(z) = A(z)^{-1} \ldots A(z+n)^{-1} v(z+n)$. For a fixed z, any $n \geq 1$ and any $s > D+2$ one finds estimates $|v(z)| \leq \frac{E}{|z+n|^{s-D-2}}$ for some constant $E > 0$. This shows that $v = 0$.

We note that the point of departure in Birkhoff's paper [9] is a difference equation of the form $y(z + 1) = z^{\mu} A_0 (1 + A_1 z^{-1} + \ldots) y(z)$ with A_0 an invertible matrix such that all the eigenvalues are distinct. Birkhoff claims to find a right lift F_{right} and a left lift F_{left} of the symbolic fundamental matrix with good asymptotic properties. The connection matrix S is defined by $F_{right}^{-1} F_{left}$ and Birkhoff proves a number of statements about S. If the eigenvalues of A_0 have absolute value 1, then Lemma 8.1 confirms this because the term z^{μ} is not essential. If the eigenvalues of A_0 do not all have absolute value 1 then some statements of [9] are no longer valid.

8.3 The connection matrix of a semi-regular equation

We begin with some comments and notations.

(1) The ring of meromorphic functions, defined on a right domain and having an asymptotic expansion in k_{∞} on the sectors at ∞ with opening $(-\pi + \epsilon, \pi + \epsilon)$ for some positive ϵ, is denoted by **Ri**. This is a difference ring containing the field of convergent Laurent series k_{∞}. One defines in a similar way a difference ring **Le** of meromorphic functions on left domains in **C**. In fact, $f(z)$ belongs to **Le** if and only if $f(-z)$ belongs to **Ri**. In other words, the functions in **Le** are defined on a left domain and have asymptotic expansion in k_{∞} on the sectors which are given, with some abuse of notation, by $(\epsilon, 2\pi - \epsilon)$ for every $\epsilon > 0$. A variation on lemma 8.1 is:

There exists a unique vector y_{left} with coordinates in **Le**, *such that the asymptotic expansion of y_{left} is $\dot{y}$ and such that $y_{left}(z + 1) = A\, y_{left}(z) + a$ holds.*

(2) An upper half plane is a subset of $\mathbf{C}$ of the form $\{z \in \mathbf{C} \mid Im(z) > c\}$ for some positive real number c. By $\mathbf{Up}$ we denote the set of meromorphic functions, defined on some upper half plane and having an asymptotic expansion in k_∞. The last statement means that this asymptotic expansion holds for every sector with $\arg(z) \in (\epsilon, \pi - \epsilon)$ with $\epsilon > 0$. The space of meromorphic functions, defined on some lower half plane and with asymptotic expansion in k_∞ is denoted by $\mathbf{Lo}$. Its definition is similar to that of $\mathbf{Up}$.

(3) Let M be a semi-regular difference module of type n over k_∞. We consider solutions of this difference equation with coordinates in various difference rings. The first ring is $k_\infty[\mu_n]$. Put

$$\omega_0(M) = ker(\Phi - 1, k_\infty[\mu_n] \otimes M).$$

The set of ϕ-invariant elements of $k_\infty[\mu_n]$ is equal to $\mathbf{C}$, because it is a subring of the universal ring introduced in Section 6.2. Hence $\omega_0(M)$ is a vector space over $\mathbf{C}$ and its dimension over $\mathbf{C}$ is equal to the dimension of M over k_∞. Moreover, the canonical map

$$k_\infty[\mu_n] \otimes_{\mathbf{C}} \omega_0(M) \to k_\infty[\mu_n] \otimes_{k_\infty} M$$

is an isomorphism.

(4) M denotes again a semi-regular difference module of type n over k_∞. We want to define $\omega_{right}(M)$, which should be the space of solutions of M living above a right domain. For this we introduce the difference ring

$$\mathbf{Ri}[\mu_n] := \mathbf{Ri}[e(c)]_{c \in \mu_n}.$$

We do not give the symbols $e(c)$, $c \in \mu_n$ an interpretation as functions on a right domain!

At this point we deviate from the current practice for difference equations. An interpretation of the symbols $e(c)$, $c \in \mu_\infty$ as functions on some domain of $\mathbf{C}$ cannot be "multiplicative" (see Section 9.2 for a discussion of this). Therefore such an interpretation does not commute with tensor products. The connection matrices, defined with such an interpretation, do not belong to the difference Galois group of the module M. Our purpose is to define connection matrices (or maps) which do belong to the difference Galois group.

The algebra $\mathbf{C}[e(c)]_{c \in \mu_n}$, also written as $\mathbf{C}[\mu_n]$, has an action of ϕ given by $\phi e(c) = c e(c)$. The group of the automorphisms of $\mathbf{C}[\mu_n]$ which commute with ϕ is denoted by C_n. This is of course a cyclic group of order n. The action of C_n is extended to a $\mathbf{Ri}$-linear action on $\mathbf{Ri}[\mu_n]$. This action commutes with the ϕ-action. Similarly one defines a C_n action on $\mathbf{Le}[\mu_n], \mathbf{Up}[\mu_n], \mathbf{Lo}[\mu_n]$. We introduce now

$$\omega_{right}(M) = ker(\Phi - 1, \mathbf{Ri}[\mu_n] \otimes_{k_\infty} M), \text{ and similarly}$$

$$\omega_{left}(M) = ker(\Phi - 1, \mathbf{Le}[\mu_n] \otimes_{k_\infty} M)$$

$$\omega_{upper}(M) = ker(\Phi - 1, \mathbf{Up}[\mu_n] \otimes_{k_\infty} M)$$

$$\text{and } \omega_{lower}(M) = ker(\Phi - 1, \mathbf{Lo}[\mu_n] \otimes_{k_\infty} M).$$

Each of these spaces has a C_n-action. There are obvious, C_n-equivariant, maps from $\omega_{right}(M)$ and $\omega_{left}(M)$ to $\omega_{upper}(M)$ and to $\omega_{lower}(M)$.

Lemma 8.2   *(1) The set of ϕ-invariant elements of $\mathbf{Ri}[\mu_n]$ is $\mathbf{C}$. The same holds for $\mathbf{Le}[\mu_n]$.*

(2) The set of ϕ-invariant elements of $\mathbf{Up}[\mu_n]$ is $\mathbf{C}\{u_n\}$ where $u = e^{2\pi i z}$ and $u_n^n = u$. The C_n-action on $\mathbf{C}\{u_n\}$ coincides with the action of the Galois group of the field extension $\mathbf{C}(\{u_n\}) \supset \mathbf{C}(\{u\})$.

(3) The set of ϕ-invariant elements of $\mathbf{Lo}[\mu_n]$ is $\mathbf{C}\{u_n^{-1}\}$ with the same action of C_n.

Proof: We consider a ϕ-invariant element f of $\mathbf{Up}[\mu_n]$. There is a unique expression $f = \sum_{c \in \mu_n} f_c e(c)$ with all $f_c \in \mathbf{Up}$. Then $\phi(f_c)c = f_c$ and so $f_c = e^{2\pi i \lambda z} h(u)$ with $\lambda \in 1/n\mathbf{Z}$, $0 \leq \lambda < 1$, $c = e^{-2\pi i \lambda}$, $u = e^{2\pi i z}$ and h a holomorphic function of u for $0 < |u| < \epsilon$ for some positive ϵ. The function f_c has an asymptotic expansion $g_c \in k_\infty$. From $\phi(g_c)c = g_c$ one concludes that g_c is zero if $c \neq 1$ and $g_r \in \mathbf{C}$ if $c = 1$. One finds the necessary and sufficient condition $h \in \mathbf{C}\{u\}$. The set of ϕ-invariant elements is therefore equal to the ring $\mathbf{C}\{u\}[t_\lambda]$, where $t_\lambda = e^{2\pi i \lambda z} e(e^{-2\pi i \lambda})$ and where λ runs over the set $[0, 1) \cap 1/n\mathbf{Z}$.

The structure of this ring is found by making a small computation. The element $t_{1/n}$ is seen to satisfy the equation $t_{1/n}^n = u$. This equation is irreducible over the field $\mathbf{C}(\{u\})$ and the subring $\mathbf{C}\{t_{1/n}\}$, of the ring of all invariants, is isomorphic to the integral closure of $\mathbf{C}\{u\}$ in the unique field extension of degree n of $\mathbf{C}(\{u\})$. With the notation $u_n = t_{1/n}$, the proof of part 2. is complete. The proof of part 3. is similar.

A ϕ-invariant element of $\mathbf{Ri}[\mu_n]$ is a sum of ϕ-invariant elements $f_c e(c)$, with $c \in \mu_n$ and $f_c \in \mathbf{Ri}$. As before, we can write $f_c = e^{2\pi i \lambda z} h(u)$. For $c = 1$, the condition that f_c has an asymptotic expansion implies that h is constant. For $c \neq 1$, the asymptotic expansion of f_c is 0 and this implies that $h = 0$. This proves the first part of 1. The proof of the second part of 1. is similar.   ∎

It is confusing to identify the element u_n, defined in the proof of Lemma 8.2, with $u^{1/n} = e^{2\pi i z/n}$. The action of ϕ is trivial on u_n, but the action of ϕ is not trivial on $u^{1/n}$! In the work of Birkhoff and others on difference equations this confusing notation is present.

Proposition 8.3 *Let M be a semi-regular difference equation of type n over k_∞.*

(1) There are canonical isomorphisms

$$\omega_0(M) \to \omega_{right}(M) \ and \ \omega_0(M) \to \omega_{left}(M).$$

(2) There are canonical isomorphisms

$$\mathbf{C}\{u_n\} \otimes_{\mathbf{C}} \omega_{right} \to \omega_{upper}(M) \ and \ \mathbf{C}\{u_n\} \otimes_{\mathbf{C}} \omega_{left} \to \omega_{upper}(M).$$

(3) There are canonical isomorphisms

$$\mathbf{C}\{u_n^{-1}\} \otimes_{\mathbf{C}} \omega_{right} \to \omega_{lower}(M) \ and \ \mathbf{C}\{u_n^{-1}\} \otimes_{\mathbf{C}} \omega_{left} \to \omega_{lower}(M).$$

Proof: 1.   Write $f \in \omega_0(M)$ as a sum $f = \sum_c f_c e(c)$ with $c \in \mu_n$ and $f_c \in k_\infty \otimes M$. The "coordinates" $\{f_c\}$ of f satisfy a semi-regular equation over k_∞. According to the Lemma 8.1 there are unique lifts $f_{right,\,c} \in \mathbf{Ri}$. The element $f_{right} := \sum_c f_{right,\,c} e(c)$ lies in $\omega_{right}(M)$. This defines the first map of part (1) of the proposition. Since the set of constants of $\mathbf{Ri}[\mu_n]$ is $\mathbf{C}$, the map is an isomorphism. We note that there is an obvious surjective homomorphism of difference rings $\mathbf{Ri}[\mu_n] \to k_\infty[\mu_n]$. Using this map, one also finds a $\mathbf{C}$-linear map $\omega_{right}(M) \to \omega_0(M)$. This is the inverse of the first map. The other part of 1. has a similar proof.

2. The map $\omega_{right}(M) \to \omega_{upper}(M)$ is induced by the obvious map of difference rings $\mathbf{Ri}[\mu_n] \to \mathbf{Up}[\mu_n]$. After taking the tensor product with $\mathbf{C}\{u_n\}$ one obtains an isomorphism since $\mathbf{C}\{u_n\}$ is the set of ϕ-invariant elements of $\mathbf{Up}[\mu_n]$. The proof of the Proposition is now clear. The term "canonical" means that the maps are functorial in M and commute with tensor products. This follows from the *uniqueness* of the asymptotic lift in Lemma 8.1.   ∎

There are now *two* isomorphisms

$$\mathbf{C}\{u_n\} \otimes_{\mathbf{C}} \omega_0(M) \to \omega_{upper}(M).$$

We define an *upper connection map* $S_{M,\,upper} \in \mathrm{Aut}(\mathbf{C}\{u_n\} \otimes_{\mathbf{C}} \omega_0(M))$ by first taking the "left" map and then the inverse of the "right" map. The definition of the lower connection map $S_{M,\,lower}$ is similar. This map has coefficients in $\mathbf{C}\{u_n^{-1}\}$. The maps $S_{M,\,upper}$ and $S_{M,\,lower}$ are C_n-equivariant. Moreover

$S_{M,\,upper}(0)$ and $S_{M,\,lower}(\infty)$ are the identity on $\omega_0(M)$.

The aim is to formulate in a precise way that a semi-regular difference module M over k_∞, of type n, is determined by the vector space $\omega_0(M)$ with its C_n-action and the two C_n-equivariant isomorphisms $S_{M,\,upper}, S_{M,\,lower}$. In particular, the difference Galois group of M should be expressed in those items. Another important point is whether one can prescribe the connection maps $S_{M,\,upper}$ and $S_{M,\,lower}$. The tool for the last question is a theorem on asymptotic expansions of matrices, due to Malgrange and Sibuya.

8.4   The theorem of Malgrange and Sibuya

In this section we discuss this theorem and adapt it for later use. Let S^1 denote the circle of directions at ∞. On this circle one considers a sheaf $\mathcal{A}$. This sheaf is defined by: $\mathcal{A}(U)$ is the ring of (germs of) meromorphic functions living on a sector corresponding with the interval U of S^1 having an asymptotic expansion in $\hat{k}_\infty$ on the sector. Let $\mathcal{A}^0$ denote the subsheaf of $\mathcal{A}$ consisting of the meromorphic functions which have asymptotic expansion 0. On the space S^1 there is an exact sequence of sheaves

$$0 \to \mathcal{A}^0 \to \mathcal{A} \to \hat{k}_\infty \to 0.$$

The last sheaf is the constant sheaf on S^1 with stalk $\hat{k}_\infty$. The exactness is a translation of the well known fact that every formal Laurent series is the asymptotic expansion of a meromorphic function on any proper sector at ∞. The cohomology sequence is

$$0 \to 0 \to k_\infty \to \hat{k}_\infty \to H^1(S^1, \mathcal{A}^0) \to \cdots$$

A special case of the theorem of Malgrange and Sibuya states that the map $\hat{k}_\infty \to H^1(S^1, \mathcal{A}^0)$ is surjective. To illustrate this, consider the two sectors S_1 and S_2 given by $\arg(z) \neq \pi$ and $\arg(z) \neq 0$. The intersection $S_1 \cap S_2$ is the union of two intervals or sectors. An element $F \in \hat{k}_\infty$ can be lifted to elements F_1, F_2 on the two sectors S_1 and S_2. Their difference $-F_1 + F_2$ lies in $\mathcal{A}^0(S_1 \cap S_2)$ and can be seen as a 1-cocycle for the group $H^1(S^1, \mathcal{A}^0)$. This describes the map $\hat{k}_\infty \to H^1(S^1, \mathcal{A}^0)$. The surjectivity of this map means the following. Let an element $f \in \mathcal{A}^0(S_1 \cap S_2)$ be given. This is a 1-cocycle representing an element ξ of $H^1(S^1, \mathcal{A}^0)$. Let $F \in \hat{k}_\infty$ have image ξ. Then the lifts F_1, F_2 on S_1, S_2 have the property that $-F_1 + F_2$ is equivalent to the 1-cocycle f. One is allowed to change F_1 and F_2 by elements in $\mathcal{A}^0(S_1)$ and $\mathcal{A}^0(S_2)$. After this change one finds an equality $-F_1 + F_2 = f$. Another formal Laurent series G with image ξ differs from F by a convergent Laurent series. The Malgrange-Sibuya theorem (see [61] for an exposition of that theorem) says that a similar result holds for

$Gl(n, \hat{k}_\infty)$ in the place of $\hat{k}_\infty$. One has again an exact sequence of non abelian groups

$$1 \to H \to Gl(n, \mathcal{A}) \to Gl(n, \hat{k}_\infty) \to 1,$$

where H denotes the subsheaf of $Gl(n, \mathcal{A})$ consisting of the matrices with asymptotic expansion 1. The corresponding cohomology sequence reads

$$1 \to 1 \to Gl(n, k_\infty) \to Gl(n, \hat{k}_\infty) \to H^1(S^1, H) \to \ldots$$

The H^1 term is not a group but only a set since it is the cohomology of a sheaf of non abelian groups. The statement of Malgrange and Sibuya is that $Gl(n, \hat{k}_\infty) \to H^1(S^1, H)$ is surjective. With the notations S_1, S_2 above this is equivalent with the following assertion:

Let $f \in H(S_1 \cap S_2)$, then there exists an element $F \in Gl(n, \hat{k}_\infty)$ and asymptotic lifts F_1, F_2 on the sectors S_1, S_2 such that $F_1^{-1} F_2 = f$. Any other $G \in Gl(n, \hat{k}_\infty)$ which maps to the same 1-cocycle has the form $G = BF$ with $B \in Gl(n, k_\infty)$.

The following statement is equivalent to the Malgrange-Sibuya theorem but does not use coordinates. Consider a triple (V, A_u, A_l) of the following data:

(a) V is a finite dimensional complex vector space.

(b) A_u is an **Up**-linear automorphism of $\mathbf{Up} \otimes V$ which has the identity on V as asymptotic expansion.

(c) A_l is a **Lo**-linear automorphism of $\mathbf{Lo} \otimes V$ which has the identity on V as asymptotic expansion.

Then there is an automorphism F of $\hat{k}_\infty \otimes V$ with F_{right} and F_{left} as asymptotic lifts. F_{right} is an automorphism of $\mathbf{Ri} \otimes V$ and F_{left} is an automorphism of $\mathbf{Le} \otimes V$, such that the "1-cocycle" $F_{right}^{-1} F_{left}$ is equal to the pair (A_u, A_l). Further F is unique up to multiplication on the left by some automorphism B of $\hat{k}_\infty \otimes V$.

Define $M \subset \hat{k}_\infty \otimes V$ as $F^{-1}(k_\infty \otimes V)$. It is clear that M is a vector space over k_∞ and that the canonical map $\hat{k}_\infty \otimes_{k_\infty} M \to \hat{k}_\infty \otimes V$ is an isomorphism. We will call such an M a k_∞-*structure* on $\hat{k}_\infty \otimes V$. Another choice BF for F does not change M. On the other hand every k_∞-structure induces a (non unique) F, automorphism of $\hat{k}_\infty \otimes V$ with $F(M) = k_\infty \otimes V$, and a unique "1-cocycle" $F_{right}^{-1} F_{left}$. In this way the collection of the triples above is in natural bijection with the k_∞-structures on $\hat{k}_\infty \otimes V$.

8.5   Regular difference equations

A regular difference module M over k_∞ is the same thing as a semi-regular module of type 1. We have associated to M the object $(\omega_0(M), S_{M,\,upper}, S_{M,\,lower})$. The first result that we will prove is:

Lemma 8.4 *Given an object* (V, T_u, T_l) *with:*

(1) V *is a complex vector space over finite dimension.*

(2) T_u *is a* $\mathbf{C}\{u\}$*-linear automorphism of* $\mathbf{C}\{u\} \otimes V$ *such that* $T_u(0)$ *is the identity on* V.

(3) T_l *is a* $\mathbf{C}\{u^{-1}\}$*-linear automorphism of* $\mathbf{C}\{u^{-1}\} \otimes V$ *such that* $T_l(\infty)$ *is the identity on* V.

Then there is a regular difference module M *over* k_∞ *and a* $\mathbf{C}$*-linear isomorphism* $f : \omega_0(M) \to V$ *such that* $f S_{M,\,upper} = T_u f$ *and* $f S_{M,\,lower} = T_l f$.

Proof: The object (V, T_u, T_l) induces an automorphism F of $\hat{k}_\infty \otimes V$ and a k_∞-structure $M \subset \hat{k}_\infty \otimes V$. The space $\hat{k}_\infty \otimes V$ is given the natural ϕ-action where ϕ is the identity on V. The ϕ-invariance of (T_u, T_l) implies that $\phi(F) = BF$ for some automorphism B of $k_\infty \otimes V$. The k_∞-structure M is invariant under ϕ. In particular M is a difference module over k_∞. It is easily seen that M has the required properties. ∎

We will now formulate in a more precise way that a regular difference equation over k_∞ is characterized by the two connection matrices. For this purpose we introduce a category $\mathcal{C}$, whose objects are the (V, T_u, T_l) with:

(1) V is a complex vector space over finite dimension.

(2) T_u is a $\mathbf{C}\{u\}$-linear automorphism of $\mathbf{C}\{u\} \otimes V$ such that $T_u(0)$ is the identity on V.

(3) T_l is a $\mathbf{C}\{u^{-1}\}$-linear automorphism of $\mathbf{C}\{u^{-1}\} \otimes V$ such that $T_l(\infty)$ is the identity on V.

A morphism $f : (V, T_u, T_l) \to (V', T'_u, T'_l)$ is a $\mathbf{C}$-linear map $f : V \to V'$ such that $f T_u = T'_u f$ and $f T_l = T'_l f$. In the obvious way one can define tensor products, (internal) Hom et cetera in $\mathcal{C}$. In this way $\mathcal{C}$ becomes a $\mathbf{C}$-linear tensor category. Let $Vect_{\mathbf{C}}$ denote the tensor category of finite dimensional complex vector spaces. The forgetful functor $\mathcal{C} \to Vect_{\mathbf{C}}$, given by $(V, T_u, T_l) \mapsto V$, is a fibre functor. Thus $\mathcal{C}$ is a neutral $\mathbf{C}$-linear Tannakian category. The following theorem contains more or less all the information about regular difference modules over k_∞.

Theorem 8.5   *(1) The functor $F : M \mapsto (\omega_0(M), S_{M, upper}, S_{M, lower})$ from the category of the regular difference modules over k_∞ to $\mathcal{C}$ is an equivalence of tensor categories.*

(2) The difference Galois group of a regular M over k_∞ is the smallest algebraic subgroup G of $Aut(\omega_0(M))$ such that $S_{M, upper} \in G(\mathbf{C}\{u\})$ and $S_{M, lower} \in G(\mathbf{C}\{u^{-1}\})$.

(3) A linear algebraic group G is the difference Galois group of a regular difference equation over k_∞ if and only if G is connected.

Proof: (1) The preceding lemma states that every object of $\mathcal{C}$ is isomorphic to some FM. This leaves us with proving that the **C**-linear map $\mathrm{Hom}(M_1, M_2) \to \mathrm{Hom}(FM_1, FM_2)$ is a bijection. It suffices to take $M_1 = \mathbf{1} :=$ the one-dimensional trivial difference module over k_∞ and $M_2 = M$ with arbitrary M. The lefthand side is equal to $\{m \in M \mid \phi(m) = m\}$. Since $F\mathbf{1} = (\mathbf{C}e, 1, 1)$, the righthand side is equal to

$$\{m \in \hat{k}_\infty \otimes M \mid \phi(m) = m, \ S_{M, upper}m = m, \ S_{M, lower}m = m\}.$$

It is clear that the two sets coincide.

(2) An object (V, T_u, T_l) of $\mathcal{C}$ generates a full tensor subcategory $\{\{(V, T_u, T_l)\}\}$ of $\mathcal{C}$. The objects of this subcategory are obtained form (V, T_U, T_l) by taking the "constructions of linear algebra", i.e. tensor products, duals, subobjects and quotient objects. The forgetful functor is again a fibre functor and the smaller category is again a neutral Tannakian category. The corresponding group G will be called *the Tannakian group of* (V, T_u, T_l). This group is actually a subgroup of $Aut(V)$ and can be described as follows: $A \in Aut(V)$ belongs to G if and only if for every construction $\mathbf{c}$ of linear algebra with result $(\mathbf{c}(V), \mathbf{c}(T_u), \mathbf{c}(T_l))$ and every subobject (V', T'_u, T'_l) the space $V' \subset \mathbf{c}(V)$ is invariant under $\mathbf{c}(A)$.
It is not difficult to see that G is actually the smallest algebraic group such that $T_u \in G(\mathbf{C}\{u\})$ and $T_l \in G(\mathbf{C}\{u^{-1}\})$. This and (1) proves (2).

(3) From (2) and $T_u(0) = 1$ and $T_l(\infty) = 1$ it follows that G is connected. On the other hand, let a connected linear group $G \subset Aut(V)$ be given. A variation on proposition 3.3 shows that there is a T_u (with $T_u(0) = 1$) such that G is the smallest algebraic group with $T_u \in G(\mathbf{C}\{u\})$. Then the object $(V, T_u, 1)$ has Tannakian group G. According to (1) this finishes the proof of (3).   ∎

Suppose now that M is a regular difference module over $\mathbf{C}(z)$. Then the right and the left lifts of elements in $\omega_0(M)$ are actually meromorphic on all of $\mathbf{C}$ (and have the required asymptotic behavior on the sectors S_1 and S_2). It follows that $S_{M, upper}$ and $S_{M, lower}$ are meromorphic on all of $\mathbf{C}$. Moreover they coincide. Put $S_M = S_{M, upper} = S_{M, lower}$. The coefficients of S_M are meromorphic functions in $u \in \mathbf{C}^*$. Further $S_M(0) = S_M(\infty) = 1$. Hence S_M has coefficients in the field $\mathbf{C}(u)$. Let us introduce another category $\mathcal{C}^*$, whose objects are the pairs (V, T) with V a finite dimensional complex vector space and T an automorphism of $\mathbf{C}(u) \otimes V$ such that $T(0) = T(\infty) = 1$. As before $\mathcal{C}^*$ is a neutral Tannakian category. In analogy with the last theorem one can prove the following:

Corollary 8.6 *(1) The functor $F : M \mapsto (\omega_0(M), S_M)$ form the category of the regular difference modules over $\mathbf{C}(z)$ to $\mathcal{C}^*$ is an equivalence of tensor categories.*

(2) The difference Galois group of a regular M over $\mathbf{C}(z)$ is the smallest algebraic subgroup G of $Aut(\omega_0(M))$ such that $S_M \in G(\mathbf{C}(u))$.

(3) A linear algebraic group G is the difference Galois group of a regular difference equation over $\mathbf{C}(z)$ if and only if G is connected.

For a matrix difference equation $y(z+1) = A\, y(z)$ (with module M) over $\mathbf{C}(z)$ one can define the *connection matrix* S as $F_{right}^{-1} F_{left}$, where F_{right} and F_{left} are the two lifts of a formal fundamental matrix F of the equation. This matrix S is clearly the matrix of the connection map S_M with respect to a given basis (depending on the choice of F) of $\omega_0(M)$. We note that the last corollary has the following consequence:

Corollary 8.7 *Let $y(z+1) = A_i y(z)$, $i = 1, 2$ denote two regular difference equations over $\mathbf{C}(z)$ with connection matrices S_i. The two difference equations are equivalent if and only if there exists a $g \in Gl(n, \mathbf{C})$ such that $S_1 = g S_2 g^{-1}$.*

8.6   Inverse problems for semi-regular equations

We want to extend the previous results for regular difference equations to the case of semi-regular modules of type n over k_∞. For this purpose we introduce a category $\mathcal{C}_n$. An object (V, T_u, T_l) of this category consists of

(1) a finite dimensional vector space V over $\mathbf{C}$;

(2) an action of C_n, the cyclic group of order n, on V;

(3) two C_n-equivariant automorphisms T_u of $\mathbf{C}\{u_n\} \otimes V$ and T_l of $\mathbf{C}\{u_n^{-1}\} \otimes V$ such that $T_u(0)$ and $T_l(\infty)$ are the identity on V.

A morphism $f : (V, T_u, T_l) \to (V', T'_u, T'_l)$ is a $\mathbf{C}$-linear map $f : V \to V'$ such that $T'_u f = f T_u$ and $T'_l f = f T_l$. The tensor product of two objects (V, T_u, T_l) and (V', T'_u, T'_l) is given as $(V \otimes V', T_u \otimes T'_u, T_l \otimes T'_l)$. The category is easily seen to be a $\mathbf{C}$-linear tensor category. Let $Vect_\mathbf{C}$ denote, as before, the tensor category of the finite dimensional vector spaces over $\mathbf{C}$. The forgetful functor $\eta : \mathcal{C}_n \to Vect_{\mathbf{C}}$, given by $(V, T_u, t_l) \mapsto V$ is a fibre functor. This makes $\mathcal{C}_n$ in to a neutral Tannakian category. The main result on the semi-regular modules of type n over k_∞ is the following.

Theorem 8.8   *(1) The functor $F : M \mapsto (\omega_0(M), S_{M, upper}, S_{M, lower})$ from the category of the semi-regular modules over k_∞ of type n to the category $\mathcal{C}_n$ is an equivalence of tensor categories.*

(2) The difference Galois group G of a semi-regular difference module M of type n over k_∞ is the smallest algebraic subgroup of $Aut(\omega_0(M))$ such that

(a) G contains the image of C_n.

(b) $S_{M, upper} \in G(\mathbf{C}\{u_n\})$.

(c) $S_{M, lower} \in G(\mathbf{C}\{u_n^{-1}\})$.

Proof: (1) It is rather clear that F is $\mathbf{C}$-linear and commutes with tensor products. In order to show that F is an equivalence we have to show that every object of $\mathcal{C}_n$ is isomorphic to some $F(M)$. Further we have to show that $\text{Hom}(M_1, M_2) \to \text{Hom}(FM_1, FM_2)$ is an isomorphism of linear vector spaces over $\mathbf{C}$.

We start with the latter and observe that it suffices to take $M_1 = \mathbf{1}$ (i.e. $\mathbf{1}$ is the trivial difference module of dimension 1) and $M_2 = M$. The first vector space is then equal to the set of elements $m \in M$ with $\phi(m) = m$. We note that $F\mathbf{1} = (\mathbf{C}e, T_u, T_l)$ with T_u and T_l the identity. The second vector space is therefore equal to the set of elements $v \in \omega_0(M)$ such that v is invariant under C_n; $S_{M, upper}v = v$ and $S_{M, lower}v = v$. For such a v the C_n-invariance implies that $v \in ker(\Phi - 1, \hat{k}_\infty \otimes M)$. The two lifts of this formal solution of the difference equation coincide on the intersection of a right domain and a left domain. It follows that $v \in M$ and $\phi(v) = v$. This proves the first part.

Let an object (V, T_u, T_l) of $\mathcal{C}_n$ be given. On the space $\mathbf{C}[\mu_n] \otimes V$ one defines a ϕ-action by the given action on $\mathbf{C}[\mu_n]$ and the trivial ϕ-action on V. The C_n-action is induced by the canonical C_n-action on $\mathbf{C}[\mu_n]$ and the given action on V. The two actions commute. Put $W := (\mathbf{C}[\mu_n] \otimes V)^{C_n}$. This is a complex vector space with trivial C_n-action and (in general) a non trivial ϕ-action. The canonical isomorphism $\mathbf{C}[\mu_n] \otimes W \to \mathbf{C}[\mu_n] \otimes V$ is equivariant for C_n and ϕ.

The automorphism T_u of $\mathbf{C}\{u_n\} \otimes V$ is extended to a $\mathbf{C}\{u_n\}[\mu_n]$-linear automorphism of $\mathbf{C}\{u_n\}[\mu_n] \otimes V = \mathbf{C}\{u_n\}[\mu_n] \otimes W$. This extension is again denoted by T_u and is equivariant for ϕ and C_n. A small calculation shows that the set of the C_n-invariant elements of $\mathbf{C}\{u_n\}[\mu_n]$ is equal to $\mathbf{C}\{e^{2\pi i z/n}\}$. This has as consequence that the set of C_n-invariant elements of $\mathbf{C}\{u_n\}[\mu_n] \otimes V$ equals $\mathbf{C}\{e^{2\pi i z/n}\} \otimes W$. The automorphism induced by T_u is denoted by S_u and is ϕ-equivariant. In a similar way one defines a ϕ-equivariant automorphism S_l of $\mathbf{C}\{e^{-2\pi i z/n}\} \otimes W$.

We apply the theorem of Malgrange and Sibuya to (W, S_u, S_l) and we find a k_∞-structure M on $k_\infty \otimes W$. Since S_u and S_l are ϕ-equivariant the k_∞-module M is invariant under ϕ, defined on $\hat{k}_\infty \otimes W$ by the ordinary ϕ on the first factor and the action of ϕ on W introduced above. Thus M is a difference module over k_∞. It is an exercise to verify that FM is isomorphic to (V, T_u, T_l).

(2) The proof is analogous to the proof given in theorem 8.5.   ∎

This theorem has an analogue for semi-regular difference modules of type n over $\mathbf{C}(z)$. Let M be such a module, then the two connection maps $S_{M, upper}$ and $S_{M, lower}$ are meromorphic on all of $\mathbf{C}$ and coincide. Put $S_M = S_{M, upper} = S_{M, lower}$. This is an automorphism of $\mathbf{C}(u_n) \otimes V$ such that $S_M(0)$ and $S_M(\infty)$ are the identity on V. Let $\mathcal{C}_n^*$ denote the category with as objects the pairs (V, T) with V a finite dimensional complex vector space and T a C_n-equivariant automorphism of $\mathbf{C}(u_n) \otimes V$ such that $T(0)$ and $T(\infty)$ are the identity on V.

Theorem 8.9 *(1) The functor $F : M \mapsto (\omega_0(M), S_M)$ from the category of the semi-regular modules over $\mathbf{C}(z)$ of type n to the category $\mathcal{C}_n^*$ is an equivalence of tensor categories.*

(2) The difference Galois group G of a semi-regular difference module M over $\mathbf{C}(z)$ is the smallest algebraic subgroup of $\mathrm{Aut}(\omega_0(M))$ such that

(a) G contains the image of C_n.

(b) $S_M \in G(\mathbf{C}(u_n))$.

The difference Galois group G of a semi-regular equation of type n over k_∞ has the property that G/G^o is cyclic of an order dividing n. Indeed, let Z denote the smallest algebraic subgroup of $\mathrm{Aut}(\omega_0(M))$ such that $S_{M, upper} \in Z(\mathbf{C}\{u_n\})$ and $S_{M, lower} \in Z(\mathbf{C}\{u_n^{-1}\})$. From the condition $S_{M, upper}(0) = S_{M, lower}(\infty) = 1$ it follows that Z is connected. The C_n-equivariance of the pair $S_{M, upper}, S_{M, lower}$ implies that the image of C_n in $\mathrm{Aut}(\omega_0(M))$ lies in the normalizer of Z. The group Z has a finite index in the group Z' generated by Z and the image of C_n. This index is a divisor of n. Hence Z' is algebraic and coincides with G. And so $Z = G^o$.

For a semi-regular difference module M of type n over $\mathbf{C}(z)$ with difference Galois group G, the situation is similar. The group G^o is the smallest algebraic group such that $S_M \in G^o(\mathbf{C}(u_n))$.

Theorem 8.10 *A linear algebraic group G is the difference Galois group of a semi-regular difference equation over k_∞ if and only if G/G^o is cyclic.*

Proof: One implication is already discussed above. Let $G \subset \mathrm{Aut}(V)$ be given such that G/G^0 is cyclic. It suffices to construct an object (V, T_u, T_l) of $\mathcal{C}_n$ with Tannakian group G.

Choose an element $d \in G$ which maps to a generator of G/G^o. The unipotent element d_u in the Jordan decomposition $d_{ss}d_u$ of d belongs to G^o. Thus d_{ss} is also mapped to a generator of G/G^o and we may (and will) suppose that d is semi-simple.

The commutative algebraic group Z generated by d is a subgroup of G. The group Z^o is a torus and Z/Z^o is cyclic. By construction, the map $Z/Z^o \to G/G^o$ is surjective. There is an element $h \in Z \subset G$, which has roots of unity as eigenvalues and maps to a generator of Z/Z^o. Thus h maps also to a generator of G/G^o.

Let the order of h be n. Then the C_n-action on V is given by the homomorphism $\rho : C_n \to \mathrm{Aut}(V)$, which sends a generator σ of C_n to $h \in G$. We will further take $T_l = 1$. The choice of the automorphism $T := T_u$ of $\mathbf{C}\{u_n\} \otimes V$ must be such that

(a) $T(0) = 1$ and G^o is the smallest algebraic group with $T \in G^o(\mathbf{C}\{u_n\})$.

(b) T is C_n-equivariant.

Condition (b) is a rather complicated one. The action of the generator σ of C_n on $\mathbf{C}\{u_n\} \otimes V$ is given by $\sigma(f \otimes v) = (\sigma(f)) \otimes h(v)$. The equivariance of T reads $T(\sigma \otimes h) = (\sigma \otimes h)T$. Put $^\sigma T = \sigma T \sigma^{-1}$. This is the natural action of σ on the points of $G^o(\mathbf{C}\{u_n\})$. Then $^\sigma T = h^{-1}Th$. Let $O(G^o)$ denote the coordinate ring of the group G^o. The action of h on G^o by conjugation induces an action of h on $O(G^o)$. The equivariant element T corresponds with an equivariant homomorphism of $\mathbf{C}$-algebras $T^* : O(G^o) \to \mathbf{C}\{u_n\}$ such that the preimage of the maximal ideal of $\mathbf{C}\{u_n\}$ is the maximal ideal of the point $1 \in G^o$. We want to produce such a T^* which is also injective. The injectivety of T^* means that $T \notin Z(\mathbf{C}\{u_n\})$ for any proper algebraic subset of G^o.

Let $O_{G^o,1}$ denote the local ring of G^o at the point $1 \in G^o$. Let $t_1, \ldots t_d$ denote generators of the maximal ideal $\underline{m}$, where d be the dimension of G^o. The t_j can be chosen such that they map to eigenvectors in $\underline{m}/\underline{m}^2$ for the action of h. In other words $h(t_j) \in \zeta_n^{b_j} t_j + \underline{m}^2$ (with ζ_n a primitive nth root of unity and $0 \leq b_j < n$). Put $s_j = \sum_{k=0}^{n-1} \zeta_n^{-kb_j} h^k(t_j)$. Then the $s_1, \ldots, s_d$ are again generators for $\underline{m}$. Further $h(s_j) = \zeta_n^{b_j} s_j$ for all j. Let $O_{G^o,1}^{an}$ denote the analytic local ring of G^o at the point $1 \in G^o$. The $s_1, \ldots, s_d$ form also a basis for the maximal ideal of this local analytic ring. Therefore $O_{G^o,1}^{an} = \mathbf{C}\{s_1, \ldots, s_d\}$. There exists a homomorphism of local analytic rings $\psi : O_{G^o,1}^{an} \to \mathbf{C}\{u_n\}$ such that each element s_j is mapped into $u_n^{b_j} u\mathbf{C}\{u\}$ and such that the $\psi(s_j)$ are algebraically independent over $\mathbf{C}$. This ψ is clearly equivariant. The composition $T^* : O(G^o) \to O_{G^o,1}^{an} \xrightarrow{\psi} \mathbf{C}\{u_n\}$ is equivariant. Its kernel is 0, since the field of fractions of $O(G^o)$ has transcendence degree d over $\mathbf{C}$. Hence T^* produces the required T.   ∎

We would like to extend the last theorem to the case of semi-regular difference equations over $\mathbf{C}(z)$. A result in this direction is:

Theorem 8.11 *A linear algebraic group G is the difference Galois group of a semi-regular difference equation over $\mathbf{C}(z)$ if*

(1) G/G° is cyclic.

(2) There exists an element of finite order $h \in G$ which maps to a generator of G/G° and lies in the connected component $N(G^\circ)^\circ$ of the normalizer $N(G^\circ)$ of G°.

Proof: It suffices to construct an object (V, T) of $\mathcal{C}_n^*$ for suitable n with Tannakian group $G \subset \mathrm{Aut}(V)$. For n we take the order of $h \in G$ and the C_n-action is given by: a generator of C_n is mapped to h.

The element h is semi-simple and lies therefore in a torus contained in the normalizer of G°. We may suppose that this torus has dimension one. The C_n-action on V can then be extended to a $\mathbf{G}_m$-action $\rho : \mathbf{G}_m \to \mathrm{Aut}(V)$ with the properties: $\rho(\zeta_n) = h$ and the image $\rho(\mathbf{G}_m)$ lies in the normalizer of G°.

Let $E \in \mathrm{Aut}(\mathbf{C}(u_n) \otimes V)$ be the image under ρ of the element u_n in the group $\mathbf{G}_m(\mathbf{C}(u_n))$. Then E lies in the normalizer of the group $G^\circ(\mathbf{C}(u_n))$. Take any $S \in G^\circ(\mathbf{C}(u))$. Then $T := E^{-1}SE$ lies in $G^\circ(\mathbf{C}(u_n))$ and T is C_n-equivariant. Indeed,

$$^\sigma T = {}^\sigma E^{-1} S {}^\sigma E = \rho(\zeta_n)^{-1} E^{-1} S E \rho(\zeta_n) = h^{-1} T h.$$

We have to make a special choice for S, in order to obtain a T with the required properties. The element S is seen as a rational map from $\mathbf{P}_{\mathbf{C}}^1$ to G°. The first condition on S is that the matrix $S - 1$ has zeroes of sufficiently high order at $u = 0$ and $u = \infty$. This guarantees that $T(0) = T(\infty) = 1$. Further we take suitable points $g_1, \ldots, g_s \in G^\circ$ such that G° is generated as an algebraic group by $\{g_1, \ldots, g_s\}$. Take $a_1, \ldots, a_s$ distinct points of $\mathbf{C}^*$ and fix elements $b_j \in \mathbf{C}^*$ with $b_j^n = a_j$. Then the rational map S should satisfy: $S(a_j) = E(b_j) g_j E(b_j)^{-1}$ for $j = 1, \ldots, s$. That an S with those properties exists follows from the fact that G° is, as a variety, isomorphic to an open set of an affine space over $\mathbf{C}$.

As we have already seen T is C_n-equivariant and $T(0) = T(\infty) = 1$. Further $T(b_j) = g_j$ for $j = 1, \ldots, s$. This implies that G° is the smallest algebraic group such that $T \in G^\circ(\mathbf{C}(u_n))$. ∎

It seems that the technical condition $h \in N(G^\circ)^\circ$ is superfluous. In fact we state the following

Conjecture: *Every linear algebraic group G with G/G° cyclic is the difference Galois group of a semi-regular difference equation over $\mathbf{C}(z)$.*

A complete proof is not available at the moment. We consider the example where G^o is a torus and the conjugation by h on G^o is not the identity. It can be seen that h does not lie in $N(G^o)^o$.

Lemma 8.12 *Suppose that the linear algebraic group G has the properties: G/G^o is cyclic and that G^o is a torus. Then G is the difference Galois group of some semi-regular difference equation over $\mathbf{C}(z)$.*

Proof: As before, there is an element $h \in G$ of finite order n which maps to a generator of G/G^o. Let G be given as an algebraic subgroup of $\mathrm{Aut}(V)$. It suffices to produce an object (V, T) of C_n^* with Tannakian group G.

Let X denote the group of the characters of G^o. Then h (or C_n) acts also on X. The element T that we have to produce is an injective homomorphism $X \to \{a \in \mathbf{C}(u_n)^* \mid a(0) = a(\infty) = 1\}$, which commutes with the C_n-actions. Indeed, the condition $T(0) = T(\infty) = 1$ is equivalent to the statement that the image of T, as a homomorphism, lies in $\{a \in \mathbf{C}(u_n)^* \mid a(0) = a(\infty) = 1\}$. The injectivity of T, as a homomorphism, is equivalent to $T \notin K(\mathbf{C}(u_n))$ for any proper algebraic subgroup K of G^o.

The action of h on the group X makes X into a module over the group ring of C_n over $\mathbf{Z}$. This group ring can be written as $\mathbf{Z}[t]/(t^n - 1)$. The multiplication by t on X is by definition the action of h on X. Suppose that X can be embedded in a $\mathbf{Z}[t]/(t^n - 1)$-module Y, such that Y is a free finitely generated $\mathbf{Z}$-module and such that an injective C_n-equivariant homomorphism $T : Y \to \{a \in \mathbf{C}(u_n)^* \mid a(0) = a(\infty) = 1\}$ exists. Then the restriction of T to X has the required properties.

The vector space $\mathbf{Q} \otimes X$ is a direct sum of irreducible representations of C_n over $\mathbf{Q}$. Any such irreducible representation is a direct summand of the regular representation $\mathbf{Q}[t]/(t^n - 1)$, where the action of h coincides with the multiplication by t. It follows that X can be considered as a $\mathbf{Z}[t]/(t^n - 1)$-submodule of $(\mathbf{Z}[t]/(t^n - 1))^N$ for some $N \geq 1$. It suffices therefore to prove the lemma for the latter module.

Let us first consider the case $X = \mathbf{Z}[t]/(t^n - 1)$. Any equivariant homomorphism $\mathbf{Z}[t]/(t^n - 1) \to \mathbf{C}(u_n)^*$ is given by $1 \mapsto w, t \mapsto \sigma(w), \ldots$. The element w has to satisfy two conditions:

(a) $w(0) = w(\infty) = 1$ and

(b) $w^{a_0} \sigma(w)^{a_1} \ldots \sigma^{n-1}(w)^{a_{n-1}} = 1$ implies that $a_0 = \ldots = a_{n-1} = 0$.

The first condition is the equivalent of $T(0) = T(\infty) = 1$. The second condition is the injectivity of T.

Choose $\lambda_1, \lambda_2 \in \mathbf{C}^*$ such that $\frac{\lambda_1}{\lambda_2}$ is not a root of unity. Then $w := \frac{u_n - \lambda_1}{u_n - \lambda_2}$ satisfies (a) and (b). Thus we have proved the lemma for the torus with character group $\mathbf{Z}[t]/(t^n - 1)$. Moreover we have produced an element T for this group such that the set of poles and zeroes of T lie in the union of two prescribed orbits in $\mathbf{C}^*$ under multiplication by ζ_n.

For the case $(\mathbf{Z}[t]/(t^n - 1))^N$, one defines a homomorphism T by elements $(w_1, \ldots, w_N)$ of the form: each $w_j = \frac{u_n - \lambda_1(j)}{u_n - \lambda_2(j)}$. The $\lambda_*(*)$ are chosen such that the orbits of the $2N$ elements $\lambda_1(j), \lambda_2(j)$ are distinct. The corresponding T is therefore again injective, C_n-equivariant and $T(0) = T(\infty) = 1$.   ∎

Chapter 9

Mild difference equations

9.1 Asymptotics for mild equations

The following theorem is the tool for the analytic study of mild difference equations. The theorem is a special case of a recent result of Braaksma and Faber.

Theorem 9.1 *Suppose that $y(z+1) = A\,y(z) + a$, where A and a have coefficients in a finite extension of k_∞, is a mild equation and let a formal solution $\hat{y}$ with coefficients in $\mathcal{P}$ be given. There exists a meromorphic vector y_{right}, such that:*

1. *y_{right} is defined on a right-domain.*

2. *y_{right} has $\hat{y}$ as asymptotic expansion at ∞ in a bounded sector with $\arg(z) \in (\frac{-\pi}{2}, \frac{\pi}{2} + \epsilon)$, for some positive ϵ.*

3. *$y_{right}(z+1) = A\,y_{right}(z) + a$.*

In general the asymptotic lift y_{right} is not unique.

Proof: Let A^c denote the "canonical form" of the difference equation (see Section 7.2). There is an invertible matrix $\hat{F}$ with coefficients in $\mathcal{P}$ such that $\hat{F}(z+1)^{-1}AF(z) = A^c$. The equation $\hat{F}(z+1) = A\hat{F}(z)(A^c)^{-1}$ satisfied by $\hat{F}$ is a mild equation. We apply now Theorem 4.1 of [16]. We refer to Theorem 11.1 for a translation of that theorem in our notations. The case that concerns us here is part 2. of Theorem 11.1. We note that the singular directions can only have $\pm\frac{\pi}{2}$ as limit directions. Let $0 < \theta < \frac{\pi}{2}$ be such that $(0, \theta]$ does not contain a singular direction. Then the multisum F of F in the direction θ lives on a sector, containing $(\theta - \frac{\pi}{2}, \theta + \frac{\pi}{2})$. For any direction $\theta' \in (0, \theta)$ one finds the same multisum. The conclusion is that F has $\hat{F}$ as asymptotic expansion on a sector

$(-\frac{\pi}{2}, \epsilon + \frac{\pi}{2})$ for some positive ϵ. Write $\dot{y}(z) = F(z)\dot{v}(z)$. Then $\dot{v}$ is a formal solution of the equation $v(z+1) = A^c\ v(z)$. Since this is an equation in canonical form, the vector $\dot{v}$ has constant coefficients. Therefore $y(z) := F(z)\dot{v}(z)$ is a solution of our problem.   ∎

Remarks 9.2 *Some notations and comments*

(1) The homogeneous equation $y(z+1) = A\ y(z)$ can have a non-zero solution which has asymptotic expansion 0 in the prescribed sector. Therefore the y_{right} of the theorem is (in general) not unique. The solution y_{right}, given in the proof of the theorem, is the multisum of the formal solution in the direction "0^+". This choice for y_{right} is unique. It is functorial on the category of mild difference modules and commutes with tensor products. We shall see below how this special choice for y_{right} can be taken as the starting point for the definition of canonical connection matrices.

(2) The ring of meromorphic functions f, defined on a right domain and having an asymptotic expansion in $\mathcal{P}$ on a sector at ∞ with opening $(-\frac{\pi}{2}, \frac{\pi}{2} + \epsilon)$ for some positive ϵ, is denoted by $\mathcal{R}$. This is a difference ring containing the field of convergent Laurent series k_∞. One defines in a similar way a difference ring $\mathcal{L}$ of meromorphic functions on parts of $\mathbf{C}$. In fact $f(z)$ belongs to $\mathcal{L}$ if and only if $f(-z)$ belongs to $\mathcal{R}$. In other words, the functions in $\mathcal{L}$ are defined on a left domain and have asymptotic expansion in $\mathcal{P}$ on a sector which is given, with some abuse of notation, by $(\frac{\pi}{2}, \frac{3\pi}{2} + \epsilon)$. A variant of Theorem 9.1 is:

There exists a vector y_{left} with coordinates in $\mathcal{L}$, such that the asymptotic expansion of y_{left} is y and such that $y_{left}(z+1) = A\ y_{left}(z) + a$ holds.

9.2 Connection matrices of mild equations

The idea of the construction of the connection matrix has already been explained. We give here some more details. The mild difference equation $y(z+1) = A\ y(z)$ has some symbolic fundamental matrix F. The coefficients of F are combinations of formal Puiseux series and symbols $e(g), l$ with $g = c\ exp(\phi(q) - q)(1 + z^{-1})^{a_0}$ (see Lemma 7.4 part 3.). On a right domain the Puiseux series are lifted, using Theorem 9.1 and the symbols l and $e(g)$, with $g \in \mathcal{G}_{mild}$ must be given an interpretation as functions on a right domain.

This interpretation has to "commute" with tensor products. Suppose for a moment that we have such an interpretation $e(g) \mapsto f(g)$, where the $f(g)$ are meromorphic functions on right domains. Of course $f(g)$ should satisfy

$\phi(f(g)) = gf(g)$. The condition that the interpretation "commutes with tensor products" implies that $f(g_1)f(g_2) = f(g_1g_2)$. Hence $f(1) = 1$ and $f(-1)^2 = f(1) = 1$. So $f(-1) = \pm 1$ and $\phi f(-1) = f(-1)$. This contradiction shows that such an interpretation is not possible.

The group of the roots of unity, which is denoted by μ_∞, is the obstruction to the construction of a suitable interpretation of the $e(g)$'s as functions. We will go as far as possible in giving the $e(g)$ an interpretation. Some, maybe arbitrary, choices are made in this process. The effect of the choices on the construction of the connection map is not essential.

Choose a **Q**-vector space $L_0 \subset \mathbf{R}$ such that $\mathbf{Q} \oplus L_0 = \mathbf{R}$. Then every element $g \in \mathcal{G}_{mild}$ can be written in a unique way as $g_0 g_1$ with $g_0 \in \mu_\infty$ and $g_1 = e^{2\pi i a_0} exp(\phi(q) - q)(1 + z^{-1})^{a_1}$ with $a_0 \in L_0$ and $a_1 \in L$ and $q = \sum_{0 < \mu < 1} a_\mu z^\mu$. The subgroup of the elements g_1 will be denoted by $\mathcal{G}_{mild,1}$. We note that the space L can be chosen as $L_0 \oplus i\mathbf{R}$. The interpretations $e(g)_{right}$, l_{right} of $e(g)$ and l on a right domain is now defined as

$$e(g)_{right} = e^{2\pi i a_0 z} \, exp\left(\sum_{0 < \mu < 1} a_\mu z^\mu \right) e^{a_1 \log(z)} \, e(g_0) \text{ and } l_{right} = \log(z),$$

where $\log(z)$ and $z^\mu := e^{\mu \log(z)}$ are defined with the ordinary choice $\log(z) = \log(|z|) + i \arg(z)$ and $\arg(z) \in (-\pi, \pi]$. On a left domain we make the interpretation $e(g)_{left}$, l_{left} is a similar way but now using another choice for the logarithm of z, namely $\log(-z) + i\pi$. The two logarithms coincide on the upper halfplane. On the lower half plane one has $(\log(-z) + i\pi) - \log(z) = 2\pi i$.

This produces a fundamental matrix F_{right} on a right domain which uses functions and the symbols $e(c)$ with $c \in \mu_\infty$. Similarly one finds a symbolic solution F_{left} on a left domain. The expression $F_{right}^{-1} F_{left}$ uses again the symbols $e(c)$ with $c \in \mu_\infty$ and functions defined on an upper half plane and a lower half plane. Thus $F_{right}^{-1} F_{left}$ is in fact a pair of matrices, say (S_{upper}, S_{lower}). We will show that S_{upper} has coordinates in $\mathbf{C}(\{u\})[e^{2\pi \lambda z} e(e^{-2\pi i \lambda})]$, where λ runs over the set $\{\lambda \in \mathbf{Q} |\ 0 \leq \lambda < 1\}$. As explained in Chapter 8, the ring above is the algebraic closure $\overline{\mathbf{C}(\{u\})}$ of the field $\mathbf{C}(\{u\})$. We note that the action of ϕ on this algebraic closure is trivial. For this reason one can not identify $\overline{\mathbf{C}(\{u\})}$ with the field of functions $\cup_{m \geq 1} \mathbf{C}(\{u^{1/m}\})$, since on the latter the action of ϕ is not the identity!

The coefficients of S_{lower} belong to the field $\overline{\mathbf{C}(\{u^{-1}\})}$. If A happens to have coordinates in $\mathbf{C}(z)$ then the two maps S_{upper} and S_{lower} glue to a "global map" S which has coordinates in $K := \mathbf{C}(u)[e^{2\pi \lambda z} e(e^{-2\pi i \lambda})]$, where λ runs over the set $\{\lambda \in \mathbf{Q} |\ 0 \leq \lambda < 1\}$. It is not difficult to see that K is a field. This field is the maximal algebraic extension of $\mathbf{C}(u)$ which is only ramified above the points $u = 0$ and $u = \infty$. The field K is the union of fields K_m. The field K_m is the unique extension of degree m of $\mathbf{C}(u)$ which is only ramified above $u = 0$ and

$u = \infty$. The field K_m has the form $\mathbf{C}(u_m)$ with $u_m^m = u$. The Galois group of K over $\mathbf{C}(u)$ is isomorphic to the profinite completion $\hat{\mathbf{Z}}$ of $\mathbf{Z}$, i.e. $\hat{\mathbf{Z}}$ is the projective limit of all $\mathbf{Z}/m\mathbf{Z}$.

A delicate point in the construction is to find the correct canonical choices for F_{right} and F_{left}. We will avoid bases and prefer to work with difference modules instead of difference equations in matrix form.

In the sequel we will use the following difference rings of functions:

$$\mathcal{R}, \ \mathcal{L}, \ \mathcal{H}_+, \ \mathcal{H}_- .$$

The first two we have already met in Remarks 9.1. We let $\mathcal{H}_+$ be the ring of meromorphic functions defined on some upper halfplane and having an asymptotic expansion in $\mathcal{P}$ for a sector with $\arg(z) \in (\frac{\pi}{2}, \frac{\pi}{2} + \epsilon)$, and $\epsilon > 0$. Similarly $\mathcal{H}_-$ denotes the ring of meromorphic functions defined on some lower halfplane having an asymptotic expansion in $\mathcal{P}$ for a sector with $\arg(z) \in (-\frac{\pi}{2}, -\frac{\pi}{2} + \epsilon)$, and $\epsilon > 0$.

The difference ring $\mathcal{P}Mild$ is the ring introduced in Lemma 7.4, $\mathcal{P}Mild = \mathcal{P}[\{e(g)\}_{g \in \mathcal{G}_{mild}}, l]$, where $\mathcal{G}_{mild}$ denotes the subgroup of $\mathcal{G}$ consisting of the elements

$$\{c \ exp(\phi(q) - q)(1 + z^{-1})^{a_0} \mid c \in \mathbf{C}^*, \ a_0 \in L, \ q = \sum_{0 < \mu < 1} a_\mu z^\mu \}.$$

This is a difference ring of "formal expressions". The next difference ring $\mathcal{R}Mild := \mathcal{R}[e(g)_{right}, l_{right}]$ uses functions and the symbols $e(c)$ with $c \in \mu_\infty$. The definitions of $\mathcal{L}Mild$, $\mathcal{H}_+Mild$, $\mathcal{H}_-Mild$ are similar. The choice of the logarithm is not important for the last two rings. Let M be a mild difference module over $\mathbf{C}(z)$ (or over $\mathbf{C}(\{z^{-1}\})$). One defines spaces of solutions of M as follows:

$$\omega_0(M) = ker(\Phi - 1, \mathcal{P}Mild \otimes M)$$

$$\omega_{right}(M) = ker(\Phi - 1, \mathcal{R}Mild \otimes M)$$

$$\omega_{left}(M) = ker(\Phi - 1, \mathcal{L}Mild \otimes M)$$

$$\omega_{upper}(M) = ker(\Phi - 1, \mathcal{H}_+Mild \otimes M)$$

$$\omega_{lower}(M) = ker(\Phi - 1, \mathcal{H}_-Mild \otimes M)$$

The tensor products are taken over $\mathbf{C}(z)$ if M is defined over that field, otherwise they are taken over $\mathbf{C}(\{z^{-1}\})$. The comparison of these spaces is the main tool in the construction of the connection map and the resulting theorems.

Lemma 9.3 *In the following u denotes $e^{2\pi i z}$.*

1. *The set of ϕ-invariant elements of $\mathcal{H}_+ Mild$ is equal to the field $\overline{C(\{u\})}$.*

2. *The set of ϕ-invariant elements of $\mathcal{H}_- Mild$ is $\overline{C(\{u^{-1}\})}$.*

3. *The sets of ϕ-invariant elements of $\mathcal{R}Mild$ and $\mathcal{L}Mild$ are contained in the field $K = \cup_m K_m$.*

Proof: (1) Let $f = \sum_{c \in \mu_\infty} f_c e(c)$ be a ϕ-invariant element. The coefficients f_c are taken in $\mathcal{H}_+ [e(g)_{right}, l_{right}]_{g \in \mathcal{G}_{mild,1}}$. The equation $\phi(f_c)c = f_c$ has the consequence $f_c = e^{-2\pi i \lambda z} h(u)$, where $\lambda \in Q$ is taken such that $0 \leq \lambda < 1$ and $e^{2\pi i \lambda} = c$. The function h is defined on $0 < |u| < \epsilon$ for some positive ϵ. The functions in $\mathcal{H}_+ [e(g)_{right}, l_{right}]_{g \in \mathcal{G}_{mild,1}}$ can be bounded on a sector $(\frac{\pi}{2}, \frac{\pi}{2} + \epsilon)$ by $c_1 e^{c_2 |z|}$ for some positive constants c_1, c_2. From this one deduces that $h \in C(\{u\})$. This proves the inclusion of the set of the ϕ-invariant elements of $\mathcal{H}_+ Mild$ into $\overline{C(\{u\})}$. The other inclusion is shown by constructing special elements.

We recall that a choice of a Q-linear subspace $L_0 \subset R$ is made such that $L_0 \oplus Q = R$. Clearly L_0 and $m + L_0$, with any $m \in Z$, are dense subsets of R. Thus there exists a $b = b_0 + m$ with $0 < b < 1$, $b_0 \in L_0$. The element $u e^{-2\pi i b z} e(e^{2\pi i b})_{right}$ is ϕ-invariant and $u e^{-2\pi i b z}$ has asymptotic expansion 0. Thus $u e^{-2\pi i b z}$ belongs to $\mathcal{H}_+$ and $u e^{-2\pi i b z} e(e^{2\pi i b})_{right}$ is a ϕ-invariant element of $\mathcal{H}_+ Mild$. From the definition of the $e(g)_{right}$ it follows that $u e^{-2\pi i b z} e(e^{2\pi i b})_{right} = u^{-m+1}$. Clearly, also $C\{u\}$ and its integral closure $\overline{C\{u\}} = \cup_{m \geq 1} C\{u_m\}$ belong to the set of ϕ-invariant elements of $\mathcal{H}_+ Mild$. Combining this one obtains the other inclusion. This proves (1). The proof of (2) is similar.

(3) A ϕ-invariant element f of $\mathcal{R}Mild$ is defined on the whole complex plane. The restrictions to an upper and a lower half plane lie in the fields $\overline{C(\{u\})}$ and $\overline{C(\{u^{-1}\})}$. From this (3) at once. *We suspect in fact that the ϕ-invariant elements of $\mathcal{R}Mild$ and $\mathcal{L}Mild$ are constants.* ∎

We note that $w_0(M)$ is a vector space over C and that the canonical map $\mathcal{P}Mild \otimes_C w_0(M) \rightarrow \mathcal{P}Mild \otimes_{C(z)} M$ is a bijection.

Lemma 9.4 *Multisummation in the direction 0^+ induces a C-linear map $w_0(M) \rightarrow w_{right}(M)$. A basis of $w_0(M)$ over C is mapped to a basis of $w_{right}(M)$ over the ring of ϕ-invariant elements of $\mathcal{R}Mild$.*

Proof: Let $y \in w_0(M)$. The element y has uniquely the form

$$y = \sum_{g \in \mathcal{G}_{mild}, n \geq 0} \hat{y}(g, n) e(g) l^n \text{ where } \hat{y}(g, n) \in \mathcal{P}.$$

The equation $\Phi(y) = y$ leads to a set of equations for the $\dot{y}(g, n)$, namely

$$\sum_n \phi(y(g,n))(l + \log(1 + z^{-1}))^n = g^{-1} \sum_n y(g,n)l^n \text{ for each } g.$$

These equations are mild and according to Theorem 9.1 there are lifts $y(g,n)_{right} \in \mathcal{R}$ of the $\dot{y}(g,n)$ such that the $\{y(g,n)_{right}\}$ satisfy the same equations. In general the lifts are not unique! The $y(g,n)_{right}$ that we take are defined by multisummation in the direction 0^+.

The element $y_{right} := \sum_{g,n} y(g,n)_{right} e(g)_{right} l_{right}$, with the interpretation of $e(g)_{right}$ and l_{right} as explained before, is Φ-invariant and consequently lies in $\omega_{right}(M)$. This defines the map in the lemma. By our construction the module M trivializes over $\mathcal{R}[\{e(g)\}_{g \in \mathcal{G}_{mild}}, l]$ and hence also over $\mathcal{R}Mild$. A basis of $\omega_0(M)$ is therefore mapped to a basis of $\omega_{right}(M)$ over the ring of ϕ-invariant elements of $\mathcal{R}Mild$. ∎

There is also a natural map $\omega_0(M) \to \omega_{left}(M)$, where we have used the other determination $(\log(-z) + i\pi)$ of the logarithm. Using Lemma 9.4 one finds natural isomorphisms

$$\overline{\mathbf{C}(\{u\})} \otimes \omega_{right}(M) \to \omega_{upper}(M) \text{ and } \overline{\mathbf{C}(\{u\})} \otimes \omega_{left}(M) \to \omega_{upper}(M).$$

Combining with the previous maps one finds a "left" and a "right" isomorphism

$$\overline{\mathbf{C}(\{u\})} \otimes_{\mathbf{C}} \omega_0(M) \to \omega_{upper}(M).$$

We define an *upper connection map* $S_{M,\,upper} \in Gl(\overline{\mathbf{C}(\{u\})} \otimes_{\mathbf{C}} \omega_0(M))$ by first taking the "left" map and then the inverse of the "right" map. The definition of $S_{M,\,lower}$ is similar. This map has coefficients in $\overline{\mathbf{C}(\{u^{-1}\})}$.

Theorem 9.5 *Let M be a mild difference module over $\mathbf{C}(z)$.*
(1) *The upper and lower connection maps of M are the restrictions of an automorphism S_M of $K_m \otimes \omega_0(M)$ with some $m \geq 1$ and K_m the unique field extension of $\mathbf{C}(u)$ of degree m which is only ramified above the points $u = 0$ and $u = \infty$.*
(2) *If M is very mild, then S_M has coordinates in the field $\mathbf{C}(u)$. Moreover S_M and S_M^{-1} are regular (i.e. defined) at the points $u = 0$ and $u = \infty$.*
(3) *The difference Galois group of M over $\mathbf{C}(z)$ is the smallest algebraic subgroup G of $Gl(\omega_0(M)) \cong Gl(n, \mathbf{C})$ such that G contains the formal difference Galois group and $S_M \in G(K_m)$.*

Proof: (1) Since the equation is defined over $\mathbf{C}(z)$, the coordinates of the upper and the lower connection maps are combinations of functions meromorphic on all of $\mathbf{C}$ and symbols $e(\zeta)$ for some roots of unity ζ. Moreover the two connection maps coincide and we can introduce the symbol S_M for both maps. Lemma 9.3 implies that that the coordinate of S_M are in $K_m = \mathbf{C}(u_m)$, for some $m \geq 1$ and with $u_m^m = u$.

(2) For very mild equations one can make the following variation on the construction of the connection maps:

The difference rings $\mathcal{P}Mild, \mathcal{R}Mild, \ldots$ are replaced by subrings $\mathcal{P}VeryMild$, $\mathcal{R}VeryMild, \ldots$ which are given as

$$\mathcal{P}VeryMild = \mathcal{P}[\{e(g)\}_{g \in \mathcal{G}_{very\ mild}}, l],$$

$$\mathcal{R}VeryMild = \mathcal{R}[\{e(g)_{right}\}_{g \in \mathcal{G}_{very\ mild}}, l_{right}],$$

etc. The spaces $\omega_{right}(M), \omega_{left}(M), \ldots$ are now defined with respect to those subrings $\mathcal{R}VeryMild, \mathcal{L}VeryMild, \ldots$. For the "very mild case" one has the following version of Lemma 9.3:

1. The set of the ϕ-invariant elements of $\mathcal{H}_+VeryMild$ is equal to the ring $\mathbf{C}\{u\}$.

2. The set of the ϕ-invariant elements of $\mathcal{H}_-VeryMild$ is equal to the ring $\mathbf{C}\{u^{-1}\}$.

The main point is that an element h of $\mathcal{H}_+VeryMild$ can be bounded on a sector $(\frac{\pi}{2}, \frac{\pi}{2} + \epsilon)$ by $c_1 e^{c_2|z|^\mu}$ with positive constants c_1, c_2 and $0 < \mu < 1$. If h is moreover ϕ-invariant then this implies that $h \in \mathbf{C}\{u\}$.

This leads to isomorphisms

$$\mathbf{C}\{u\} \otimes_{\mathbf{C}} \omega_0(M) \to \omega_{upper}(M),$$

$$\mathbf{C}\{u^{-1}\} \otimes_{\mathbf{C}} \omega_0(M) \to \omega_{lower}(M).$$

Hence $S_{M,\,upper}$ and $S_{M,\,lower}$ are invertible maps on the spaces $\mathbf{C}\{u\} \otimes_{\mathbf{C}} \omega_0(M)$ and $\mathbf{C}\{u^{-1}\} \otimes_{\mathbf{C}} \omega_0(M)$. Hence $S_M = S_{M,\,upper} = S_{M,\,lower}$ and its inverse have coordinates in the subring of $\mathbf{C}(u)$ consisting of the rational functions which are defined at 0 and at ∞.

(3) The map S_M depends in a functorial way on M. Its construction commutes with "all constructions of linear algebra", in particular with tensor products. It follows that the difference Galois group G must satisfy $S_M \in G(K_m)$ (and of course G contains the formal difference Galois group). On the other hand, let Z denote the smallest algebraic subgroup of $Gl(\omega_0(M))$ containing the formal difference Galois group and satisfying $S_M \in Z(K_m)$. Then the set of Z-invariant elements of $\omega_0(M)$ is easily seen to be $ker(\Phi - 1, M)$. From this one can conclude that $Z = G$. ∎

A similar proof can be worked out for an analytic mild difference module, i.e. a mild difference module over k_∞.

Corollary 9.6 *Let M be a mild difference module over $k_\infty = \mathbf{C}(\{z^{-1}\})$. The upper and lower connection matrices have coordinates in the fields $\mathbf{C}(\{u\})$ and $\mathbf{C}(\{u^{-1}\})$. The difference Galois group of M is the smallest algebraic subgroup G of $Gl(\omega_0(M))$ such that:*

1. $G_{M,\,formal} \subset G$.

2. $S_{M,\,upper} \in G(\overline{\mathbf{C}(\{u\})})$.

3. $S_{M,\,lower} \in G(\overline{\mathbf{C}(\{u^{-1}\})})$.

Corollary 9.7 *Let M be a mild difference module over $\mathbf{C}(z)$ then M and $k_\infty \otimes M$ have the same difference Galois group.*

Proof: The difference Galois group G of $\hat{k}_\infty \otimes M$ is certainly contained in the one of M. From $S_{M,\,upper} \in G(\overline{\mathbf{C}(\{u\})})$ it follows that $S_M \in G(K)$ since $S_M = S_{M,\,upper}$. According to 9.5, G is also the difference Galois group of M. ∎

We take a small break in order to point out a consequence of Corollary 9.7. A result of Birkhoff states that every differential module M over k_∞ comes from a differential module N over $\mathbf{C}(z)$ (i.e. M is isomorphic to $k_\infty \otimes N$). It is somewhat of a surprise that this does not hold for difference equations. We give two examples.

Example 9.8 The difference equation $y(z+1) = (1+z^{-1})^{1/m} y(z)$ is not equivalent over k_∞ with a difference equation over $\mathbf{C}(z)$. To see this consider the matrix difference equation

$$y(z+1) = \begin{pmatrix} e^{2\pi i/n} & 0 \\ 0 & (1+z^{-1})^{1/m} \end{pmatrix} y(z)$$

The formal fundamental matrix of the equation is

$$\begin{pmatrix} e(e^{2\pi i/n}) & 0 \\ 0 & e((1+z^{-1})^{1/m}) \end{pmatrix}.$$

Its difference Galois group over k_∞ (and also over $\hat{k}_\infty$) is $\mathbf{Z}/n\mathbf{Z} \times \mathbf{Z}/m\mathbf{Z}$. If a difference equation over $\mathbf{C}(z)$ is equivalent with our example then this equation is also mild and has therefore the same difference Galois group. However for a difference Galois group G over $\mathbf{C}(z)$ one knows that G/G^o is cyclic. This contradiction proves that the matrix difference equation is not equivalent to a matrix equation over $\mathbf{C}(z)$. Since the upper part of the equation is already defined over $\mathbf{C}(z)$ it follows that the lower part is not equivalent over k_∞ with an equation over $\mathbf{C}(z)$

Example 9.9 The difference equation $y(z+1) = (1+z^{-2})^{1/m}y(z)$ over k_∞ does not come from an equation over $\mathbf{C}(z)$. To see this note that the equation is regular and has an upper and a lower connection matrix, called s_{upper} and s_{lower}. For the equation $v(z+1) = (1+z^{-2})v(z)$, which is the mth tensor power of the first equation, we will show later in 10.2 that the connection matrix is $s = \frac{(u-e^{2\pi})(u-e^{-2\pi})}{(u-1)^2}$. Hence $s_{upper} = s^{1/m} = 1 + *u + *u^2 + \ldots \in \mathbf{C}\{u\}$ and a similar expression for s_{lower}. If the equation of the example came from an equation over $\mathbf{C}(z)$ then s_{upper} is the germ at $u = 0$ of a rational function in u. However s is not an mth power in $\mathbf{C}(u)$.

Corollary 9.10 *Let M be a regular singular difference module over $\mathbf{C}(z)$.*

1. *The connection matrix S_M has coordinates in $\mathbf{C}(u)$. Further $S_M(0) = 1$ and $S_M(\infty)$ is the formal monodromy map of M.*

2. *The difference Galois group of M is the smallest algebraic subgroup G of $Gl(\omega_0(M))$ such that $S_M \in G(\mathbf{C}(u))$. In particular, the difference Galois group of M is connected.*

Proof: 1. According to Theorem 9.5, the coordinates of S_M are in some $K_m = \mathbf{C}(u_m)$. The number $m \geq 1$ corresponds to the occurrence of the mth roots of unity in the formal classification of the equation. In the regular singular case $m = 1$.
Let $y \in \omega_0(M)$ have the form

$$y = \sum_{a \in \mathbf{C}/\mathbf{Z},\ n \geq 0} \hat{y}(a,n) e((1+z^{-1})^a) l^n.$$

The term $\mathbf{C}/\mathbf{Z}$ indicates $\{a \in \mathbf{C}|\ 0 \leq Re(a) < 1\}$. Then one has, with the obvious notations,

$$y_{right} = \sum_{a \in \mathbf{C}/\mathbf{Z},\ n \geq 0} y(a,n)_{right} e^{a \log z} (\log z)^n$$

$$y_{left} = \sum_{a \in \mathbf{C}/\mathbf{Z},\ n \geq 0} y(a,n)_{left} e^{a(\log(-z)+i\pi)} (\log(-z)+i\pi)^n$$

On the upper half plane $\log(z) = \log(-z) + i\pi$ and the $y(a,n)_{right}$ and $y(a,n)_{left}$ have the same asymptotic expansion. This implies $S_M(0) = 1$. On the lower half plane $\log(-z) + i\pi = \log(z) + 2\pi i$ and the formula for y_{left} can be written there as

$$y_{left} = \sum_{a \in \mathbf{C}/\mathbf{Z},\, n \geq 0} y(a,n)_{left} e^{2\pi i a} e^{a \log(z)} (\log(z) + 2\pi i)^n.$$

From this the statement about $S_M(\infty)$ follows.

2. The formal difference Galois group is easily seen to be generated by the formal monodromy map. From Theorem 9.5 and $S_M(\infty) =$ the formal monodromy group, the first statement follows. The group G is connected since $S_M(0) = 1$. ∎

Remark 9.11 *For an analytic regular singular difference module M, i.e. M is defined over $k_\infty = \mathbf{C}(\{z^{-1}\})$, Corollary 9.10 has an obvious analogue:*

1. *$S_{M,\,upper}(0) = 1$ and $S_{M,\,lower}(\infty)$ is the formal monodromy map.*

2. *The difference Galois group is the smallest algebraic group G such that $S_{M,\,upper} \in G(\mathbf{C}\{u\})$ and $S_{M,\,lower} \in G(\mathbf{C}\{u^{-1}\})$.*

Let d be a complex number and $y(z+1) = A(z)y(z)$ a difference equation. *The shift by d of this equation is the equation $y(z+1) = A(z+d)y(z)$. Note that $y(z)$ satisfies $y(z+1) = A(z)y(z)$ if and only if $\tilde{y}(z) = y(z+d)$ satisfies $\tilde{y}(z+1) = A(z+d)\tilde{y}(z)$* Let M denote the module corresponding to the matrix $A(z)$, then $\tau_d M$ will denote the module corresponding to the matrix $A(z+d)$.

Lemma 9.12 *Let $S = S(u_m)$ denote the connection matrix (total, upper or lower) of the mild difference module M, then $S(e^{2\pi i d/m} u_m)$ is the connection matrix of the shifted module $\tau_d M$.*

Proof: We will quickly verify that the shift τ_d can be defined on all the objects that are needed for the construction of the connection matrix of a mild module and that τ_d commutes with the construction of the connection matrix. Indeed, τ_d acts on $\mathcal{P}$ by $\tau_d(z^\lambda) := (1 + dz^{-1})^\lambda z^\lambda$. Further $\tau_d(l) := l + \log(1 + dz^{-1})$. For $g = c \, exp(\phi(q) - q)(1 + z^{-1})^{a_0} \in \mathcal{G}_{mild}$ one defines

$$\tau(g) = \left(\frac{1 + (d+1)/z}{(1 + d/z)(1 + 1/z)} \right)^{a_0} exp(\phi(\tau_d(q) - q) - (\tau_d(q) - q))g.$$

We note that g is multiplied under the action of τ_d by a convergent Puiseux series since $\tau_d(q) - q$ is a Puiseux series in positive powers of z^{-1}. The action of τ_d on $\mathcal{P}Mild$ commutes with the action of ϕ. One defines the action of τ_d on $\mathcal{R}, \mathcal{L}, \mathcal{H}_+, \mathcal{H}_-$ in the natural way, i.e. $(\tau_d f)(z) = f(z+d)$. Next one can verify that the natural action of τ_d on the functions $e(g)_{right}, e(g)_{left}, l_{right}, l_{left}$ is compatible with the τ_d action on the symbols $e(g), l$ for $g \in \mathcal{G}_{mild}$. This defines an action of τ_d on $\mathcal{R}Mild$ et cetera, still commuting with ϕ. Finally, the unicity of the multisummation proves that $y_{right}(z+d)$ coincides with the lift of the the shifted formal solution of the shifted difference equation. ∎

9.3 Tame differential modules

In order to obtain nice formulas for connection matrices we extend the construction of the connection matrix to *tame difference modules*. A difference module M over k_∞ (or over $C(z)$) will be called tame (over k_∞) if M can be written as a finite direct sum

$$M = \sum_{\lambda \in \mathbf{Z}} k_\infty e(z^\lambda) \otimes M_\lambda, \text{ where}$$

each M_λ is a mild difference module over k_∞. This decomposition is unique because the modules M_λ are supposed to be mild.

One can extend this notion as follows: M is *tame over the finite field extension* $k_\infty(z^{1/m})$ *of* k_∞ if $k_\infty(z^{1/m}) \otimes M$ can be written as a finite direct sum

$$k_\infty(z^{1/m}) \otimes M = \sum_{\lambda \in 1/m\mathbf{Z}} k_\infty(z^{1/m})e(z^\lambda) \otimes M_\lambda, \text{ where}$$

each M_λ is a mild difference module over $k_\infty(z^{1/m})$.

The decomposition is again unique.

The module is tame (over $k_\infty(z^{1/m})$) if and only if M can be represented by a matrix difference equation $y(z+1) = A\, y(z)$ (over $k_\infty(z^{1/m})$) such that A is a direct sum of blocks. Each block has the form $z^\lambda(B_0 + B_{1/m}z^{-1/m} + \ldots)$ with $\lambda \in \mathbf{Q}$ and B_0 an invertible matrix.

For a tame module M we want to define the connection matrix. The space of formal solutions $\omega_0(M)$ is the kernel of $\Phi - 1$ acting on $\mathcal{P}[\{e(g)\}_{g\in\mathcal{G}}, l] \otimes M$. We have to make a decision how to lift the symbol $e(z)$ to a elements $e(z)_{right}$ and $e(z)_{left}$. In the literature one finds more than one choice. Here we will take the same choice as G.D. Birkhoff in [9], namely:

$$e(z)_{right} = \Gamma(z) \text{ and } e(z)_{left} = -2\pi i e^{i\pi z}\Gamma(1-z)^{-1}.$$

Reasons for this choice are the following. The Γ-function has no zeros on $C \setminus (-\infty, 0]$ and is then the exponential of a unique holomorphic $F(z)$ on the same domain with $F(1) = 0$. Thus we can define $e(z^\lambda)_{right}$ as $e^{\lambda F(z)}$. Similarly $e^{i\pi z}\Gamma(1-z)^{-1}$ has no zeros on $C \setminus [1, \infty)$ and is the exponential of a holomorphic function G on the same domain with $G(0) = 0$. Then $e(z^\lambda)_{left}$ will be $e^{\lambda G(z)}$. The constant in the expression for $e(z)_{left}$ are chosen such that the "connection matrix" of the equation $y(z+1) = zy(z)$ has a simple form. This connection matrix is $e(z)_{right}^{-1}e(z)_{left} = 1 - u$.

The natural definition of the connection matrix S of a tame difference module $M = \sum_\lambda k_\infty e(z^\lambda) \otimes M_\lambda$ is the direct sum of the connection matrices of the pieces. The connection matrix of $k_\infty e(z^\lambda) \otimes M_\lambda$ is the product of the connection matrix of M_λ with the constant $(1-u)^\lambda$. The meaning of $(1-u)^\lambda$ is defined by the choices of F and G above. In particular $(1-u)^\lambda$ has value 1 for $u = 0$.

Finally we note that connection matrices for tame modules respect tensor products.

9.4   Inverse problems for mild equations

In the first part of this section we will study equivalence classes of mild difference equations. This theme returns, in a more complicated setting, for wild difference equations. The last part gives a construction of a mild difference equation over k_∞ with a prescribed difference Galois group.

Fix a *mild* difference equation $y(z+1) = S\, y(z)$ over the algebraic closure of k_∞. We consider the set of all difference equations $y(z+1) = A\, y(z)$ which are formally equivalent, i.e. over the field $\mathcal{P}$, to $y(z+1) = S\, y(z)$. As we know, the formal isomorphism $\hat{F}$ with $\hat{F}(z+1)^{-1}A\hat{F}(z) = S$ can be lifted to F_{right} on an "extended right half plane", i.e. on a sector of the form $-\frac{\pi}{2} < \arg(z) < \frac{\pi}{2} + \delta$ for some $\delta > 0$. There is also a lift F_{left} on an "extended left half plane", i.e. on a sector of the form

$$\{z \in \mathbf{C} \mid z = |z|e^{i\phi},\ |z| > R,\ \frac{\pi}{2} < \phi < \frac{3\pi}{2} + \delta\}.$$

The positive number δ depends only on S.

The matrix $T := F_{left}^{-1}F_{right}$ has the properties:

1. T is defined on the union of two sectors $\frac{\pi}{2} < \arg(z) < \frac{\pi}{2} + \delta$ and $-\frac{\pi}{2} < \arg(z) < -\frac{\pi}{2} + \delta$.

2. T is asymptotically the identity on this union of the two sectors, i.e. this property holds on any proper subsector defined by $\leq$ signs.

3. $T(z+1)^{-1}ST(z) = S$.

We will call a matrix T with the properties (1),(2) and (3) above a *connection cocycle*. To the pair $(A, \hat{F})$ we associate the connection cocycle $F_{left}^{-1}F_{right}$. This connection cocycle is not unique, since the lifts F_{right} and F_{left} are not unique. Another choice of those two matrices changes T into $U_{left}TU_{right}$ where U_{right} is asymptotically the identity on the extended right half plane and $U_{right}(z+1)^{-1}SU_{right}(z) = S$. The other matrix U_{left} has similar properties. We call T and $U_{left}TU_{right}$ with the stated properties equivalent. Let $H^1(S)$ denote the set of equivalence classes of the matrices T.

On the set of pairs $(A, \hat{F})$ satisfying $\hat{F}(z+1)^{-1} A \hat{F}(z) = S$ we introduce also an equivalence relation. The pairs $(A_i, \hat{F}_i)$, $i = 1, 2$ are called equivalent if there is an invertible meromorphic matrix C, defined over the algebraic closure of k_∞, such that $C(z+1)^{-1} A_1 C(z) = A_2$ and $\hat{F}_1 = C \hat{F}_2$. Equivalent pairs produce equivalent connection cocycles.

Theorem 9.13 *The map from the set of equivalent classes of pairs $(A, \hat{F})$, satisfying $\hat{F}(z+1)^{-1} A \hat{F}(z) = S$, to $H^1(S)$ is bijective.*

Proof: The verification of the injectivity of the map is a straightforward matter. We will prove now that the map is surjective. Let T be a connection cocycle. The Theorem of Malgrange-Sibuya (see section 8.4) asserts that there is an invertible formal matrix $\hat{F}$ (with coefficients in $\mathcal{P}$) and two asymptotic lifts F_1, F_2 of $\hat{F}$ defined on sectors $(-\frac{\pi}{2}, \frac{\pi}{2} + \delta)$ and $(\frac{\pi}{2}, \frac{3\pi}{2} + \delta)$ (with some abuse of notation) such that $F_1^{-1} F_2 = T$. The two invertible meromorphic matrices $A_i = F_i(z+1) S F_i(z)^{-1}$, $i = 1, 2$ are defined on the two sectors above. On the intersection of the two sectors one has $A_1 = A_2$. Hence the pair A_1, A_2 glues to an invertible meromorphic matrix A with coefficients in the algebraic closure of k_∞. The connection cocycle is obviously the image of the pair $(A, \hat{F})$. ∎

Corollary 9.14 *Suppose, in addition to the theorem, that the coefficients of S are in $\mathbf{C}(z)$ and suppose that the connection cocycle T extends to an invertible meromorphic matrix on all of $\mathbf{C}$. Then the connection cocycle is the image of a pair $(A, \hat{F})$ such that the coefficients of A are in $\mathbf{C}(z)$.*

Proof: We use the notation of the proof of the last theorem. Let the pair F_1, F_2 satisfy $F_1^{-1} F_2 = T$. Then clearly F_1 and F_2 are invertible meromorphic matrices on all of $\mathbf{C}$. The same thing holds for $A = F_1(z+1) S F_1(z)^{-1} = F_2(z+1) S F_2(z)^{-1}$. Since A is also invertible meromorphic at ∞ one has that the entries of A are in $\mathbf{C}(z)$. ∎

Remark 9.15

The multisummation theorem associates to a pair $(A, \hat{F})$ with $\hat{F}(z+1)^{-1} A \hat{F} = S$ *unique* asymptotic lifts F_{right} and F_{left} on an extended right half plane and an extended left half plane. Thus multisummation provides a connection matrix $T = F_{left}^{-1} F_{right}$ which is a representative of the equivalence class of the connection cocycle in $H^1(S)$ of $(A, \hat{F})$.

Theorem 9.16 *Every linear algebraic group G which has a finite commutative subgroup K with at most two generators and such that $K \to G/G^\circ$ is surjective is the difference Galois group of a mild difference equation $y(z+1) = A\, y(z)$ over the field k_∞.*

Proof: In principle we will use Corollary 9.6 in order to obtain the required difference equation. However this corollary is not very constructive and we will in fact work with a subclass $\mathcal{M}_{m,n}$ of the mild difference equations for which the connection matrices (with a slightly different definition) and the formal monodromy can be prescribed.

For this subclass we will use the method of the proof of Theorem 8.10. Fix two integers $n, m \geq 1$. A mild difference module M over k_∞ belongs to this subclass if $\hat{k}_\infty[e(e^{-2\pi i/m}), e((1+z^{-1})^{-1/n})] \otimes M$ is a trivial difference module. In other words, there exists a fundamental matrix for M with coordinates in the difference ring $\hat{k}_\infty[e(e^{-2\pi i/m}), (e((1+z^{-1})^{-1/n})]$.

The group of the automorphisms (commuting with ϕ and $\hat{k}_\infty$-linear) of this difference ring is denoted by H. It is the product of two cyclic groups of order m and n. The two generators σ_1, σ_2 are given by

$$\sigma_1 e(e^{-2\pi i/m}) = e^{2\pi i/m} e(e^{-2\pi i/m}); \ \sigma_1 e((1+z^{-1})^{-1/n}) = e((1+z^{-1})^{-1/n}),$$

$$\sigma_2 e(e^{-2\pi i/m}) = e(e^{-2\pi i/m}); \ \sigma_2 e((1+z^{-1})^{-1/n}) = e^{2\pi i/n} e((1+z^{-1})^{-1/n}).$$

This group will act on the constructions that we will make.

We introduce difference rings $\mathcal{R}', \mathcal{L}', \mathcal{H}'_+, \mathcal{H}'_-$ as the subrings of $\mathcal{R}, \mathcal{L}, \mathcal{H}_+, \mathcal{H}_-$ having their asymptotic expansion in $\hat{k}_\infty$. The spaces $\omega_0(M), \omega_{right}(M), \dots$ are now defined as

$$ker(\Phi - 1, \hat{k}_\infty[e(e^{-2\pi i/m}), e((1+z^{-1})^{-1/n})] \otimes M),$$

$$ker(\Phi - 1, \mathcal{R}'[e(e^{-2\pi i/m}), e((1+z^{-1})^{-1/n})] \otimes M),$$

etc. We note that we keep the symbol $e((1+z^{-1})^{-1/n})$ and will not replace it by, say $e((1+z^{-1})^{-1/n})_{right}$. On each of the space $\omega_0(M), \omega_{right}(M), \dots$ there is an action of H. As in the general method for mild equations we obtain two connection maps, again denoted by $S_{M, \, upper}, S_{M, \, lower}$. They are not the same as before since we have not given the symbol $e((1+z^{-1})^{-1/n})$ an interpretation. The two connection maps are H-equivariant. Moreover $S_{M, \, upper}(0) = 1, S_{M, \, lower}(\infty) = 1$. The upper connection map is an invertible map with coordinates in the ring of the ϕ-invariant elements of $\mathcal{H}'_+[e(e^{-2\pi i/m}), e((1+z^{-1})^{-1/n})]$.

We will need this ring explicitly. A ϕ-invariant element is a sum of expressions of the form

$$e^{2\pi i z \frac{a}{m}} z^{\frac{b}{n}} h_{a,b}(u) e(e^{-2\pi i a/m}) e((1+z^{-1})^{-b/n}) \text{ with } 0 \leq a < m, 0 \leq b < n$$

and $h_{a,b}(u)$ a meromorphic function of u is some set $0 < |u| < \epsilon$. The condition that $e^{2\pi i z \frac{a}{m}} z^{\frac{b}{n}} h_{a,b}(u)$ has an asymptotic expansion in $\hat{k}_\infty$ implies that $h_{a,b}$ lies either in $\mathbf{C}\{u\}$ or $u\mathbf{C}\{u\}$. Put $F_1 = e^{2\pi i z/m} e(e^{-2\pi i/m})$ (this was called u_m earlier) and $F_2 = z^{1/n} e((1 + z^{-1})^{-1/n})$. We note that $F_2^n = 1$. Then the ring of the ϕ-invariant elements is a subring of $\mathbf{C}\{u\}[F_1, F_2]$. It is in fact the subring $\mathbf{C}\{u\}[F_1, uF_2, uF_2^2, \dots uF_2^{n-1}]$. This is an analytic local ring of dimension one over $\mathbf{C}$.

A similar calculation can be done for the determination of the ring of ϕ-invariant elements $(\mathcal{H}'_-[e(e^{-2\pi i/m}), e((1 + z^{-1})^{-1/n})])^\phi$. For notational convenience we denote the two rings of ϕ-invariants by A_{upper} and A_{lower}.

We have now associated to M in the subclass $\mathcal{M}_{m,n}$ the following object $(V, T_u, T_l) = (\omega_0(M), S_{M, upper}, S_{M, lower})$ with:

(a) V a finite dimensional vector space over $\mathbf{C}$ with an H-action.

(b) T_u is an H-equivariant automorphism of $A_{upper} \otimes V$ such that $T_u(0) = 1$.

(c) Similar statement for T_l.

As in the proof of Theorem 8.8, one can show that the subclass $\mathcal{M}_{m,n}$ is equivalent to the category of all triples (V, T_u, T_l) as above. The Tannakian group of a triple is the smallest group G which contains the action of H on V and satisfies $T_u \in G(A_{upper}), T_l \in G(A_{lower})$. The connected component G^o is in fact the smallest algebraic group with $T_u \in G^o(A_{upper}), T_l \in G^o(A_{lower})$. The map $H \to G \to G/G^o$ is surjective.

Let now $G \in GL(V)$ satisfying the conditions of the theorem be given. Choose suitable m, n and a surjective $H \to K$. Then we should produce a triple (V, T_u, T_l) such that G^o is the smallest algebraic group with $T_u \in G^o(A_{upper}), T_l \in G^o(A_{lower})$. We start by taking $T_l = 1$. Further T_u corresponds to an H-equivariant homomorphism of $\mathbf{C}$-algebras $\psi' : O(G^o) \to A_{upper}$ such that the preimage of the maximal ideal of A_{upper} is the maximal ideal of the point $1 \in G^o$. It suffices to produce a ψ' which is also injective, since in that case $T_u \notin Z(A_{upper})$ for any proper closed subset of G^o.

Let $O_{G^o, 1}$ denote the local ring of G^o at $1 \in G^o$. Let d be the dimension of G^o. There is a basis $t_1, \dots, t_d$ of the maximal ideal of $O_{G^o, 1}$ such that each t_j is an eigenvector for the action of H. In other words, there are $0 \le a_j < m$ and $0 \le b_j < n$ such that $\sigma_1 t_j = e^{2\pi i a_j/m} t_j; \sigma_2 t_j = e^{2\pi i b_j/n} t_j$. The analytic local ring of G^o at 1 is denoted by $O_{G^o, 1}^{an}$. This ring has also $t_1, \dots t_d$ as generators for its maximal ideal and so $O_{G^o, 1}^{an} = \mathbf{C}\{t_1, \dots, t_d\}$. An H-equivariant homomorphism of local analytic rings $\psi : O_{G^o, 1}^{an} \to A_{upper} \subset \mathbf{C}\{u\}[F_1, F_2]$ is now described by: each t_j is mapped to an element of $u^\epsilon F_1^{a_j} F_2^{b_j} \mathbf{C}\{u\}$ (with ϵ equal to 0 or 1). It is clear that one can choose the images $\psi(t_j)$ such that they are algebraically independent over $\mathbf{C}$. This implies that $\psi' : O(G^0) \to O_{G^o, 1}^{an} \xrightarrow{\psi} A_{upper}$ is injective.

∎

Remark 9.17

The condition in Theorem 9.16, imposed on the group G, is also necessary for the group to be a difference Galois group of any difference equation over k_∞. This will be proved in Proposition 10.11.2.

Chapter 10

Examples of equations and Galois groups

10.1 Calculating connection matrices

For a regular difference module M over $\mathbf{C}(z)$ or k_∞ we will give more or less explicit formulas for the connection matrices. A regular module has a presentation in matrix form $y(z + 1) = A\ y(z)$ with $A = 1 + A_2 z^{-2} + \ldots$. The formal fundamental matrix F has the form $1 + F_1 z^{-1} + \ldots$. The right and left lifts of F is denoted by F_{right} and F_{left}. The connection matrix is $F_{right}^{-1} F_{left}$ and, in general, is defined for $z \in \mathbf{C}$ with $|Im(z)| \gg 0$. Its restriction to $\{z \in \mathbf{C}|\ Im(z) \gg 0\}$ is the matrix of $S_{M,\,upper}$ on a suitable basis of $\omega_0(M)$. Similarly for $S_{M,\,lower}$. If M is defined over $\mathbf{C}(z)$ then $A = 1 + A_2 z^{-2} + \ldots$ is chosen with coefficients in $\mathbf{C}(z)$. In that case $F_{right}^{-1} F_{left}$ is defined on all of $\mathbf{C}$ and is equal to the matrix of S_M with respect to a suitable basis of $\omega_0(M)$.

Proposition 10.1 *Let M be a regular difference module over k_∞ or over $\mathbf{C}(z)$. Let a matrix equation $y(z + 1) = A\ y(z)$, representing M, be chosen such that $A = 1 + A_2 z^{-2} + A_3 z^{-3} + \ldots$ and A has rational coefficients if M is defined over $\mathbf{C}(z)$. The formal fundamental matrix F has the form $1 + F_1 z^{-1} + \ldots$ and its lifts F_{right} and F_{left} are equal to*

$$\lim_{n \to \infty} A(z)^{-1} A(z + 1)^{-1} \cdots A(z + n - 1)^{-1} \ and$$

$$\lim_{n \to \infty} A(z - 1) A(z - 2) \cdots A(z - n + 1).$$

The connection matrix $F_{right}^{-1} F_{left}$ is equal to

$$\lim_{n \to \infty} A(z + n - 1) \cdots A(z) A(z - 1) \cdots A(z - n + 1).$$

Proof: From $F_{right}(z+1) = A(z)F_{right}(z)$ it follows that

$$F_{right}(z) = A(z)^{-1}A(z+1)^{-1}\cdots A(z+n-1)^{-1}F_{right}(z+n).$$

It is easily seen that $\lim_{n\to\infty} A(z)^{-1}\cdots A(z+n-1)^{-1}$ converges (locally) uniformly on a right-domain. Further $\lim_{n\to\infty} F_{right}(z+n)$ converges (locally) uniformly to 1. This proves the first statement. The second statement can be obtained by replacing z by $-z$. The third assertion is now obvious. ∎

Example 10.2 *The equation $y(z+1) = a(z)y(z)$ with $a(z) \in \mathbf{C}(z)^*$.*

Every order one equation $y(z+1) = ay(z)$ over $\mathbf{C}(z)$ is tame and defines a connection matrix which is an element in $\mathbf{C}(u)^*$. We will denote the induced map $\mathbf{C}(z)^* \to \mathbf{C}(u)^*$ by $\mathcal{C}$. Since the connection matrix respects tensor products, $\mathcal{C}$ is a homomorphism of groups. Let $\mathbf{C}(z)^*_{mild}$ denote the elements of $\mathbf{C}(z)^*$ which have value $\neq 0, \infty$ at ∞. Every element of $\mathbf{C}(z)^*$ has uniquely the form $z^n m$ with $n \in \mathbf{Z}$ and m mild. By definition $\mathcal{C}(z^n m) = (1-u)^n \mathcal{C}(m)$. We want to prove that $(\mathcal{C}f)(e^{2\pi id}u) = \mathcal{C}(\tau_d f)(u)$. For mild elements this has been proved in Lemma 9.12. Hence we have only to verify this property for z. This amounts to proving that

Lemma 10.3 $\mathcal{C}(1 + dz^{-1}) = \dfrac{1 - e^{2\pi id}u}{1-u}.$

Proof: Write $(1 + d/z) = (1 + z^{-1})^d b$. The element $b \in \mathbf{C}\{z^{-1}\}$ is regular. The symbolic solution f of the equation $y(z+1) = (1 + d/z)y(z)$ has lifts $f_{right} = e^{d\log(z)}\prod_{n=0}^{\infty} b(z+n)^{-1}$ and $f_{left} = e^{-a(\log(-z)+i\pi)}\prod_{n=1}^{\infty} b(z-n)^{-1}$. Then f_{right} has an analytic continuation on all of $\mathbf{C}$. Using the formula for f_{right} and the equation $f_{right}(z+1) = (1 - az^{-1})f_{right}(z)$ one finds that f_{right} has poles of order 1 in a subset of $a + \mathbf{Z}$ and has zeros of order 1 in a subset of $\mathbf{Z}$. For f_{left} one finds poles of order 1 in a subset of $\mathbf{Z}$ and zeros of order 1 in a subset of $a + \mathbf{Z}$. The connection matrix $s = f_{right}^{-1} f_{left}$ has as function of u a pole of order 1 for $u = 1$ and a zero of order 1 for $u = e^{2\pi id}$. The condition $s(0) = 1$ leads then to the required formula $s = \frac{1 - e^{2\pi id}u}{1-u}$. ∎

Now that we know that $\mathcal{C}$ has the correct behavior with respect to shifts and is a homomorphism one can easily prove the formula

$$cz^n \prod_j (z - c_j)^{n_j} \overset{\mathcal{C}}{\mapsto} (1-u)^n \prod_j (1 - e^{-2\pi ic_j}u)^{n_j}, \text{ where } c, \, c_j \in \mathbf{C}^*.$$

In other terms, the connection matrix of $a := cz^n \prod_j (z - c_j)^{n_j}$, with divisor $n[0] - n[\infty] + \sum n_j[c_j]$, is the unique element $s \in \mathbf{C}(u)^*$ with $s(0) = 1$ and with divisor $n[1] - n[\infty] + \sum n_j[e^{2\pi ic_j}]$.

Remarks 10.4

(1) First we make some comments on the proof that the connection matrix of $(1 + d/z)$ is $s = \frac{1-e^{2\pi i d}u}{1-u}$. The evaluation $s(\infty) = e^{2\pi i d}$ is (as it should be) the formal monodromy. Further the connection matrix $s \in \mathbf{C}(u)^*$ has the form $s_{upper} = \prod_{n=-\infty}^{\infty} b(z + n)$ and $s_{lower} = e^{2\pi i d} \prod_{n=-\infty}^{\infty} b(z + n)$. This shows that the product $\prod_{n=-\infty}^{\infty} b(z + n)$ converges on the upper halfplane to $\frac{1-e^{2\pi i d}u}{1-u}$ and on the lower halfplane to $e^{-2\pi i d}\frac{1-e^{-2\pi i a}u}{1-u}$.

(2) According to 2.31 the difference Galois group of the equation $y(z+1) = ay(z)$ with $a \in \mathbf{C}(z)^*$ over $\mathbf{C}(z)$ is equal to the cyclic group of order m if and only if a has the form $\zeta\frac{f(z+1)}{f(z)}$ for some primitive mth root of unity ζ and some $f \in \mathbf{C}(z)$. One can reformulate this as follows:

The equation has a difference Galois group of order $m \geq 1$, if and only if:

(*) $a(\infty)$ is a primitive mth root of unity and

(**) The restriction of the divisor of a to any $\mathbf{Z}$-orbit in $\mathbf{C}$ (i.e. a set of the form $c + \mathbf{Z} \subset \mathbf{C}$) has degree 0.

The formula for the connection matrix s of the equation $y(z + 1) = ay(z)$ with $a = cz^n \prod_j (z - c_j)^{n_j}$ with c, $c_j \in \mathbf{C}^*$ shows that $s = 1$ if and only if the divisor $\sum n_j[e^{2\pi i c_j}]$ is trivial and $n = 0$. This is precisely the same as $a(\infty) \in \mathbf{C}^*$ and the condition (**) above

(3) Every element $t \in \mathbf{C}(u)^*$ with $t(0) = t(\infty) = 1$ is the image under $\mathcal{C}$ of a regular element $a \in \mathbf{C}(z)^*$, i.e. with $a(\infty) = 1$. This is a special case of Corollary 8.6.

(4) We have seen that the connection matrix of the equation $y(z + 1) = z^\lambda y(z)$, which was produced in a formal way, "commutes" with the shift operators τ_d. This implies that the connection map of any *tame* differential module also "commutes" with the shifts τ_d.

Example 10.5 *The equation $y(z + 1) - y(z) = a$ with $a \in \mathbf{C}(z)$.*

Put $A = \begin{pmatrix} 1 & a \\ 0 & 1 \end{pmatrix}$ with $a \in \mathbf{C}(z)$. The equation $y(z+1) = A\,y(z)$ is very mild. Indeed, put $B := \begin{pmatrix} 1 & b \\ 0 & 1 \end{pmatrix}$ with a suitable $b \in \mathbf{C}[z]$. Then

$B(z + 1)^{-1}A(z)B(z)$ is equal to $\begin{pmatrix} 1 & \tilde{a} \\ 0 & 1 \end{pmatrix}$ with $\tilde{a} \in \mathbf{C}(z)$ such that $\tilde{a}(\infty) = 0$.

The symbolic fundamental matrix F has the form $\begin{pmatrix} 1 & f \\ 0 & 1 \end{pmatrix}$ with $f \in \mathbf{C}((z^{-1})) + \mathbf{C}\, l$ satisfying $f(z+1) - f(z) = a$. The connection matrix reads $\begin{pmatrix} 1 & b \\ 0 & 1 \end{pmatrix}$, with $b \in \mathbf{C}(u)$ such that $b(0) = 0$. The connection matrix defines a map $\mathbf{C}(z) \to \mathbf{C}(u)$ which will be denoted by $\mathcal{L}$. We will derive an explicit formula $\mathbf{C}$-linear map $\mathcal{L}$. The key for this is the formula:

Lemma 10.6 $\mathcal{L}(z^{-1}) = \dfrac{-2\pi i u}{1 - u}.$

Proof: Write $z^{-1} = b + \log(1 + z^{-1})$. Then $b \in z^{-2}\mathbf{C}\{z^{-1}\}$. The symbolic solution f is equal to $l + B$ with $B \in z^{-1}\mathbf{C}[[z^{-1}]]$. Then $f_{right} = \log(z) - \sum_{n=0}^{\infty} b(z+n)$ and $f_{left} = \log(-z) + i\pi + \sum_{n=1}^{\infty} b(z-n)$. The function f_{right} has an analytic continuation to $\mathbf{C}$ and has poles of order 1 in a subset of the negative integers. Similarly f_{left} has poles of order 1 in a subset of the positive integers. $\mathcal{L}(z^{-1})$ is the difference of those two functions and has a pole of order 1 for $u = 1$ and a zero for $u = 0$. Hence $\mathcal{L}(z^{-1}) = \frac{cu}{1-u}$ for some constant c. The expression $f_{left} - f_{right}$ is on the lower halfplane equal to $2\pi i + \sum_{n=-\infty}^{\infty} b(z+n)$. For $z \to -i\infty$ the infinite sum tends to zero. This proves that $c = -2\pi i$. ∎

It can be seen that $\mathcal{L}$ has the property $\mathcal{L}(\frac{d}{dz}a) = (2\pi i u)\frac{d}{du}(\mathcal{L}(a))$. With this relation one derives the formula

$$\mathcal{L}(z^{-k}) = \frac{(-2\pi i)^{k+1}}{(k-1)!} \sum_{n=1}^{\infty} n^k u^n.$$

Using that $\mathcal{L}$ "commutes" with shifts one finds a similar formula for $\mathcal{L}((z+d)^{-k})$. Using those formulas for $\mathcal{L}$ one can show the following:

1. Every $b \in \mathbf{C}(u)$ with $b(0) = b(\infty) = 0$ is the image of an element $a \in \mathbf{C}(z)$ with $\mathrm{ord}_{\infty}(a) \geq 2$.

2. $\mathcal{L}(a) = 0$ if and only if there is a $f \in \mathbf{C}(z)$ with $f(z+1) - f(z) = a$.

The last statement has in fact two proofs. The first one observes that $f(z+1) - f(z) = a$ has a solution in $\mathbf{C}(z)$ if and only if a is the sum (over n and α) of expressions of the form:

$$\sum_{j \in \mathbf{Z}} \frac{\beta_j}{(z - \alpha + j)^n} \quad \text{with} \quad \sum_j \beta_j = 0.$$

A calculation shows that $\mathcal{L}(a) = 0$ leads precisely to this formula for a.

The second proof starts with the statement that $\mathcal{L}(a) = 0$ is equivalent to f (i.e. the element in the formal fundamental matrix) lies in $\mathbf{C}\{z^{-1}\}$. The equation $f(z+1) - f(z) = a$ implies that f is then in $\mathbf{C}(z)$.

Example 10.7 *The equation* $y(z+1) = e^{2\pi i/m} y(z) + a$ *with* $a \in \mathbf{C}(z)$.

Let us write $\zeta_m = e^{2\pi i/m}$ and $e_m = e(\zeta_m)$. We suppose that a is chosen such that the equation has a formal solution f which is divergent. The choice $a = z^{-1}$ produces such an f. The equation reads in matrix form

$$y(z+1) = \begin{pmatrix} \zeta_m & a \\ 0 & 1 \end{pmatrix} y(z).$$

A symbolic fundamental matrix is

$$\begin{pmatrix} 1 & f \\ 0 & 1 \end{pmatrix} \begin{pmatrix} e_m & 0 \\ 0 & 1 \end{pmatrix}.$$

The two lifts $\begin{pmatrix} 1 & f_{right} \\ 0 & 1 \end{pmatrix}$ and $\begin{pmatrix} 1 & f_{left} \\ 0 & 1 \end{pmatrix}$ produce a connection matrix $\begin{pmatrix} 1 & g \\ 0 & 1 \end{pmatrix}$ with $g = e_m^{m-1}(f_{left} - f_{right})$. The mth power of g is an element $h \in \mathbf{C}(u)$. In general (for instance for $a = z^{-1}$) the equation $X^m - h$ is irreducible over $\mathbf{C}(u)$. This example shows that the finite extensions $\mathbf{C}(u_m)$ with $u_m^m = u$ are needed for the description of the connection matrix. ∎

Proposition 10.8 *Let* $y(z+1) = A \, y(z)$ *be a regular singular equation with coefficients in* $\mathbf{C}(z)$. *Then the equation is equivalent (over* $\mathbf{C}(z)$) *to an equation where* A *has the form*

$$\frac{z^e}{(z-a_1)\ldots(z-a_e)}(A_0 + A_1 z^{-1} + \ldots + A_e z^{-e}),$$

with $a_i \in \mathbf{C}^*$ *and* $A_0 = 1$.
The connection matrix S *of the equation is*

$$S = \frac{u^e}{\prod_j (1 - e^{-2\pi i a_j} u)}(B_0 + B_1 u^{-1} + \ldots + B_e u^{-e}),$$

with B_0 *invertible and* $B_e = 1$.

Proof: The first statement follows easily form the definition of regular singular. Let F_{right} and F_{left} denote the two fundamental matrices and so $S = F_{right}^{-1} F_{left}$. The matrix F_{right} is an invertible holomorphic matrix for z with $Re(z) \gg 0$.

Similarly for F_{left}. Put $G_{right} := \Gamma(z - a_1) \ldots \Gamma(z - a_e) F_{right}$. This is an invertible holomorphic matrix for z with $Re(z) \gg 0$. Define $G_{left} := \prod_j \{(-2\pi i e^{i\pi(z-a_j)}) \Gamma(1 - (z - a_j))^{-1}\} F_{left}$. Then G_{left} is holomorphic for z with $Re(z) \ll 0$. Both G_{right} and G_{left} satisfy the difference equation $G_*(z + 1) = (A_0 z^e + \ldots A_{e-1} z + A_e) G_*(z)$. This implies that G_{left} has no poles in $\mathbf{C}$. The periodic matrix $G_{right}^{-1} G_{left} = \prod_j (1 - e^{-2\pi i a_j} u) S$ has for z with $Re(z) \gg 0$ no poles. Thus this expression is a matrix polynomial in the variable u. From $S(0) = 1$ and $S(\infty)$ is an invertible constant matrix one concludes that $\prod_j (1 - e^{-2\pi a_j} u) S$ has the form $u^e B_0 + u^{e-1} B_1 + \ldots + B_e$. Moreover $B_e = S(0) = 1$ and $B_0 = S(\infty)$ is invertible.   ∎

We note that a somewhat similar statement is present in [9]. Birkhoff's *fundamental problem* is to show that the map, which associates to $(A_0 = 1, \ldots, A_e)$ the $(B_0, \ldots, B_{e-1}, B_e = 1)$, is surjective. The fundamental problem asks in fact for the possibilities of the connection map. In Corollary 8.6 a complete answer is given for the case of regular difference equations over $\mathbf{C}(z)$. The answer, translated in the formulation of Birkhoff reads as follows:

Let $B_* := (B_0, B_1, \ldots, B_{e-1}, B_e)$ *with* $B_e = B_0 = 1$ *be given. Then there is a* $(A_0, \ldots, A_e)$, *with* $A_0 = 1$ *and* $A_1 = -(a_1 + \ldots + a_e)id$, *which has image* B_*.

One can extend this to the case of regular singular difference equations over $\mathbf{C}(z)$. Let $\mathcal{T}$ denote the neutral Tannakian category, for which the objects are the pairs (V, T) with V a finite dimensional vector space over $\mathbf{C}$ and T an automorphism of $\mathbf{C}(u) \otimes_{\mathbf{C}} V$ such that $T(0) = 1$ and $T(\infty)$ is an invertible constant map. In our terminology the answer to Birkhoff's question can be stated by the next result:

The morphism from the category of the regular singular difference modules over $\mathbf{C}(z)$ *to the category* $\mathcal{T}$ *is an equivalence of tensor categories. In particular, every matrix* $T \in Gl(n, \mathbf{C}(u))$ *with* $T(0) = 1$ *and* $T(\infty)$ *is an invertible matrix, is the connection matrix of some regular singular matrix equation* $y(z+1) = A\, y(z)$ *with* $A \in Gl(n, \mathbf{C}(z))$.

The proof of Corollary 8.6 for the regular situation can be extended to the regular singular case. We will not give the details.

10.2 Classification of order one equations

Let k be a difference field. The order one equation $\phi(y) = ay$ is equivalent to $\phi(y) = by$ if and only if there is an $f \in k^*$ with $ab^{-1} = \frac{\phi(f)}{f}$. Let $U(k)$ denote the subgroup of k^* consisting of the elements $\frac{\phi(f)}{f}$, with $f \in k^*$. The group of equivalence classes of order one equations over k is therefore equal to $k^*/U(k)$.

In this section we will consider for k the fields k_∞ and $\hat{k}_\infty$ with $\phi(z) = z + 1$. An easy calculation shows that $U(\hat{k}_\infty)$ consists of the power series $1 + kz^{-1} + a_2 z^{-2} + a_3 z^{-3} + \ldots$, where $k \in \mathbf{Z}$ and $a_2, a_3, \ldots \in \mathbf{C}$. Every order one difference equation is therefore formally equivalent to a unique equation

$$y(z + 1) = z^n c(1 + dz^{-1})y(z) \text{ with } n \in \mathbf{Z}, \; c \in \mathbf{C}^*, \; d \in \mathbf{C},$$

such that $0 \leq Re(d) < 1$.

We will show that the classification of difference equations over k_∞ is quite a different affair. This is unlike the (local) classification of order one differential equations, where the formal and the analytic theory coincide.

Let the group H (with additive group law) be defined by the exactness of the following sequence:

$$0 \to H \to k^*_\infty/U(k_\infty) \to \hat{k}^*_\infty/U(\hat{k}_\infty) \to 1.$$

We introduce the following notations:

- S^1 is the unit circle, seen as the circle of directions at ∞.

- $\mathcal{A}^0_{per}$ denotes the sheaf on S^1 of germs of 1-periodic meromorphic functions which are *flat at* ∞. One defines this as follows: Let $(a, b) \subset S^1$ be an interval, then an element ξ of $\mathcal{A}^0_{per}(a, b)$ is the germ of a meromorphic function f defined on a bounded sector at ∞ with $\arg(z) \in (a, b)$ such that $f(z + 1) = f(z)$ and f has 0 as asymptotic expansion on this sector.

The following proposition describes the difference between analytic and formal classification of order one equations at ∞.

Proposition 10.9 *H is isomorphic to the complex vector space $H^1(S^1, \mathcal{A}^0_{per})$. The complex vector space H is infinite dimensional.*

Proof: An element h of H is represented by some $a \in k^*_\infty$ such that $f(z + 1) = af(z)$ has a solution $f \in \hat{k}^*_\infty$. Write $f = cz^k g$ with $c \in \mathbf{C}^*$, $k \in \mathbf{Z}$ and $g = 1 + g_1 z^{-1} + g_2 z^{-2} + \ldots$. Then h can also be represented by $b := \frac{g(z+1)}{g(z)} = 1 + b_2 z^{-2} + b_3 z^{-3} + \ldots \in k^*_\infty$. One writes $b = exp(B)$ and $g = exp(G)$, with $B = B_2 z^{-2} + B_3 z^{-3} \ldots \in z^{-2}\mathbf{C}\{z^{-1}\}$ and $G = G_1 z^{-1} + G_2 z^{-2} + \ldots \in z^{-1}\mathbf{C}[[z^{-1}]]$. The relation between B and G is the formula $(\phi - 1)G = B$. A small calculation shows that the map $(\phi - 1) : z^{-1}\mathbf{C}[[z^{-1}]] \to z^{-2}\mathbf{C}[[z^{-1}]]$ is bijective. It follows that H can be identified with the cokernel of the map

$$(\phi - 1) : z^{-1}\mathbf{C}\{z^{-1}\} \to z^{-2}\mathbf{C}\{z^{-1}\}.$$

Let B and G be as above. According to Lemma 8.1, the formal power series G can be lifted to a unique meromorphic G_{right} satisfying:

1. G_{right} lives on a sector at ∞ with $\arg(z) \in (-\pi, \pi)$ and has G as asymptotic expansion on this sector. Let $S_1 \subset S^1$ denote the interval $(-\pi, \pi)$ of the circle of directions S^1 at ∞.

2. $G_{right}(z+1) - G_{right}(z) = B(z)$ holds in this sector.

A similar asymptotic lift G_{left} exists on a sector at ∞, given by $\arg(z) \neq 0$. Let $S_2 \subset S^1$ denote the interval of the circle of directions at ∞ corresponding with this sector. With an abuse of notation S_2 is the interval $(0, 2\pi)$. The difference $G_{right} - G_{left}$ is 1-periodic and has asymptotic expansion 0 on $S_1 \cap S_2$. Hence

$$G_{right} - G_{left} \in \mathcal{A}^0_{per}(S_1 \cap S_2).$$

This element is seen as a 1-cocycle for the sheaf $\mathcal{A}^0_{per}$ on S^1 with respect to the covering $\{S_1, S_2\}$ of S^1. Let $\alpha(h)$ denote the image of this 1-cocycle in the first cohomology group of the sheaf. Thus we have defined an additive map

$$\alpha : H \to H^1(S^1, \mathcal{A}^0_{per}).$$

The sheaf $\mathcal{A}^0_{per}$ on S^1 is of a rather special nature:

1. On the interval $\arg(z) \in (0, \pi)$, the sheaf is constant with as stalk the complex vector space consisting of all holomorphic functions $\sum_{n=1}^{\infty} a_n e^{2\pi i n z}$ converging on some upper half plane. Write again $u = e^{2\pi i z}$. Then this condition is equivalent to $\sum_{n=1}^{\infty} a_n u^n \in u\mathbf{C}\{u\}$.

2. On the interval $\arg(z) \in (\pi, 2\pi)$ (with abuse of notation), the sheaf is constant with as stalk the expressions $\sum_{n=-1}^{-\infty} a_n e^{2\pi i n z}$ which converge on some lower halfplane. In other terms, $\mathcal{A}^0_{per}$ is on $(\pi, 2\pi)$ the constant sheaf with stalk $u^{-1}\mathbf{C}\{u^{-1}\}$.

3. $\mathcal{A}^0_{per}(S) = 0$ if the interval $S \subset S^1$ contains one or both the direction $\arg(z) = 0$ or the direction $\arg(z) = \pi$.

On the intervals S_1 and S_2 and on their intersection $S_1 \cap S_2$, the sheaf $\mathcal{A}^0_{per}$ has trivial cohomology. The H^1 of the sheaf is therefore equal to the Čech-cohomology with respect to the covering $\{S_1, S_2\}$ of S^1. Thus

$$H^1(S^1, \mathcal{A}^0_{per}) = \mathcal{A}^0_{per}(S_1 \cap S_2),$$

which is certainly an infinite dimensional vector space over $\mathbf{C}$.

Finally, we have to show that $\alpha : H \to H^1(S^1, \mathcal{A}^0_{per})$ is bijective. If the 1-cocycle $G_{right} - G_{left}$, corresponding to $h \in H$ is trivial, then $G_{right} = G_{left}$

and thus $G \in z^{-1}\mathbf{C}\{z^{-1}\}$. This implies $h = 0$. Hence α is injective.

In order to prove the surjectivity of α we first consider the sheaf $\mathcal{A}^0$ on S^1. This is the sheaf (of germs) of meromorphic functions on sectors at ∞ having asymptotic expansion 0. According to a result of Malgrange and Sibuya (see also Chapter 8), there is a natural isomorphism

$\mathbf{C}[[z^{-1}]]/\mathbf{C}\{z^{-1}\} \to H^1(S^1, \mathcal{A}^0)$. This isomorphism is induced by the map

$\beta : \mathbf{C}[[z^{-1}]] \to H^1(S^1, \mathcal{A}^0)$ defined by $\beta(F) = (F_{right} - F_{left})$,

where F_{right}, F_{left} are meromorphic functions on the sectors S_1 and S_2 with asymptotic expansion $F \in \mathbf{C}[[z^{-1}]]$. The difference $(F_{right} - F_{left})$ is a 1-cocycle for the sheaf $\mathcal{A}^0$ with respect to the covering $\{S_1, S_2\}$ of S^1.

Let a 1-cocycle $f \in \mathcal{A}^0(S_1 \cap S_2)$ be given. Then there is a $F \in \mathbf{C}[[z^{-1}]]$ with $\beta(F) = F_{right} - Fleft$ is equivalent to f. One can change F_{right} and F_{left} by flat functions such that actually $F_{right} - F_{left} = f$. For a fixed choice of F the F_{right} and F_{left} are now uniquely determined by the condition $F_{right} - F_{left} = f$. The F itself is unique up to a change $F + H$ with $H \in \mathbf{C}\{z^{-1}\}$. The lifts of $F + H$ are now $F_{right} + H$ and $F_{left} + G$ and the satisfy again $(F_{right} + H) - (F_{left} + H) = f$.

Let a 1-cocycle f for $\mathcal{A}^0_{per}$ be given. Since $\mathcal{A}^0_{per}$ is a subsheaf of $\mathcal{A}^0$, there is a formal power series G, which we may suppose to lie in $z^{-1}\mathbf{C}[[z^{-1}]]$, and there are lifts G_{right}, G_{left} such that $f = G_{right} - G_{left}$. Then $G(z + 1)$ with lifts $G_{right}(z + 1), G_{left}(z + 1)$ also satisfies $G_{right}(z + 1) - G_{left}(z + 1) = f$. Hence $G(z + 1) - G(z) = h \in z^{-2}\mathbf{C}\{z^{-1}\}$. It is clear from the construction that $\alpha(h)$ is equal to the 1-cocycle f. $\blacksquare$

We remark Proposition 10.9 is related to Corollary 8.6 and to the calculations in Example 10.5.

10.3 More on difference Galois groups

In Example 9.8 we have encountered an equation with difference Galois group $\mathbf{Z}/n\mathbf{Z} \times \mathbf{Z}/m\mathbf{Z}$ (both over k_∞ and $\hat{k}_\infty$). The first factor comes from the decomposition of the Picard-Vessiot ring as a product of domains and the second factor comes from the finite field extension of k_∞ (or $\hat{k}_\infty$) present in the Picard-Vessiot ring.

We can make this more precise by a further analysis of the automorphism group $Aut(R/\hat{k}_\infty, \phi)$ of the universal Picard-Vessiot ring $R = \mathcal{P}[\{e(g)\}_{g \in \mathcal{G}}, l]$. We will use (for the moment) the notation $\mathbf{G}$ for this group of automorphisms. As noted in Chapter 6, $\mathbf{G}$ has the structure of an affine group scheme over $\mathbf{C}$. For any difference module M (over $\mathbf{C}(z)$, k_∞ or $\hat{k}_\infty$) the group $\mathbf{G}$ acts on

$\omega_0(M) = ker(\Phi - 1, R \otimes M)$. The image of this action is the formal difference Galois group, i.e. the difference Galois group of the module $\hat{k}_\infty \otimes M$. We introduce now two elements γ and δ of $\mathbf{G}$. Their images in the formal difference Galois group $G_{M, formal}$ of any M will be denoted by γ_M and δ_M. The choice for γ and δ is made in such a way that the images γ_M and δ_M have finite order in $G_{M, formal}$ and that their images in $G_{M, formal}/G^o_{M, formal}$ commute and generate this group.

The first element γ is almost the same as the formal monodromy. We define the action of γ on $\mathcal{P}$ by $\gamma(z^a) = e^{2\pi i a} z^a$ for all $a \in \mathbf{Q}$. The action of γ on R is determined by $\gamma(e(g)) = e(\gamma g)$ and $\gamma(l) = l$. This last choice makes γ distinct from the formal monodromy. The Picard-Vessiot ring of any difference module M over $\hat{k}_\infty$ is a subring of R and is generated over $\hat{k}_\infty$ by a finite collection of elements: z^a with $a \in \mathbf{Q}$; finitely many elements $e(g)$, with $g \in \mathcal{G}$; and possibly l. From this is it clear that the image γ_M in the formal difference Galois group of M has finite order.

For the definition of the second element $\delta \in \mathbf{G}$, we need a homomorphism $\epsilon : \mathbf{C}^* \to \mu_\infty$, where μ_∞ denotes the subgroup of $\mathbf{C}^*$ consisting of the roots of unity. We require that ϵ is the identity on μ_∞. Such a homomorphism can be given by choosing, as before, a $\mathbf{Q}$-linear subspace L of $\mathbf{C}$ such that $\mathbf{C} = \mathbf{Q} \oplus L$. The map ϵ is then defined by $\epsilon(e^{2\pi i(\lambda + a)}) = e^{2\pi i \lambda}$ for any $\lambda \in \mathbf{Q}$ and any $a \in L$. The second automorphism δ acts as the identity on $\mathcal{P}[l]$ and $\delta(e(g)) = h(g)e(g)$ with $h : \mathcal{G} \to \mathbf{C}^*$ defined by $h(z^\lambda c(1 + z^{-1})^{a_0} exp(\phi(q) - q)) = \epsilon(c)$. It is easily seen that the image of δ in any formal difference Galois group has finite order.

Let us introduce a subgroup $\mathbf{G}^o$ of $\mathbf{G}$ as follows: $\mathbf{G}^o$ consists of the automorphisms of $R/\hat{k}_\infty$ (commuting with ϕ), which are the identity on $\mathcal{P}[\{e(c)\}]_{c \in \mu_\infty}$. This group scheme $\mathbf{G}^o$ has the property that its image in the formal difference Galois group $G_{M, formal}$ of any difference module coincides with its connected component of the identity $G^o_{M, formal}$. The elements γ and δ do not commute. The commutator $\alpha := \delta\gamma\delta^{-1}\gamma^{-1}$ is the identity on the subring $\mathcal{P}[\{e(g)\}_{g \in \mathcal{G}_{mild}}, l]$ and $\alpha e(z^\lambda) = e^{2\pi i\lambda} e(z^\lambda)$. In particular, $\alpha \in \mathbf{G}^o$. Hence the image of α in any formal difference Galois group $G_{M, formal}$ lies in the connected component $G^o_{M, formal}$. Moreover, if M is mild then the $e(z^\lambda)$ (for $\lambda \neq 0$) are not present in the Picard-Vessiot ring of M and as a consequence the images γ_M and δ_M commute. From the explicit form of R and the γ, δ we come now to the following results on the formal difference Galois group $G_{M, formal}$ of a difference module M over $\hat{k}_\infty$:

Proposition 10.10 *Let K denote the subgroup of $G_{M, formal}$ generated by the images γ_M and δ_M of γ and δ. Then:*

1. *The map $K \to G_{M, formal}/G^o_{M, formal}$ is surjective and the group $G_{M, formal}/G^o_{M, formal}$ is commutative and has at most two generators.*

2. *If M is mild, then the subgroup $K \subset G_{M, formal}$ is a finite commutative group with at most two generators.*

Let M denote a difference module over k_∞ or $\mathbf{C}(z)$. Let the Picard-Vessiot ring for M be denoted by PV and that of $\hat{k}_\infty \otimes M$ by $\hat{PV}$. There is an injective morphism $f : PV \to \hat{PV}$. This morphism is not unique, since it can be composed with a ϕ-invariant automorphism of $\hat{PV}$. However the image of f does not depend on the choice of f. For a fixed choice of f one finds an inclusion of the formal difference Galois group $G_{M, formal} \to G$, where G is the difference Galois group of M. This inclusion is unique up to conjugation by an element of G. In the sequel we will fix f and the inclusion $G_{M, formal} \to G$. The elements γ_M and δ_M are now also considered as elements of G. Likewise, the subgroup K of $G_{M, formal}$ is considered as a subgroup of G.

Proposition 10.11   1. *Let M be a difference module over $\mathbf{C}(z)$ with difference Galois group G. Then G/G° is generated by the image of δ_M.*

2. *Let M be a difference module over k_∞ with difference Galois group G. Then the images of δ_M and γ_M in G/G° commute and generate this group.*

3. *Suppose that M is a mild difference module over k_∞. Then the subgroup K of G generated by γ_M and δ_M is commutative and finite. Moreover $K \to G/G^\circ$ is surjective.*

Proof: 1. As above, the Picard-Vessiot ring of M over $\mathbf{C}(z)$ is denoted by PV. The ring PV is embedded in the Picard-Vessiot ring $\hat{PV}$ of $\hat{k}_\infty \otimes M$ over $\hat{k}_\infty$. Let $G' \subset G$ denote the subgroup of G generated by G° and δ_M. The inclusion $G_{M, formal} \subset G$ implies $G^0_{M, formal} \subset G^\circ$. Let $G'_{M, formal}$ denote the subgroup of $G_{M, formal}$ generated by $G^\circ_{M, formal}$ and δ_M. Then $G'_{M, formal} \subset G'$. Then for the rings of invariants one has the following inclusion $PV^{G'} \subset \hat{PV}^{G'_{M, formal}}$. The last ring of invariants is a finite field extension of $\hat{k}_\infty$. Since G' has finite index in G one has that $PV^{G'}$ is a finite extension of $\mathbf{C}(z)$. This finite extension lies in a finite extension of $\hat{k}_\infty$ and is therefore a field. The extension $PV^{G'} \supset \mathbf{C}(z)$ is a finite extension of difference fields. This implies that $PV^{G'}$ is equal to $\mathbf{C}(z)$. By Lemma 1.31, $G = G'$.

2. The Picard-Vessiot ring of M over k_∞ is denoted by PV. The one of $\hat{k}_\infty \otimes M$ by $\hat{PV}$. One fixes an inclusion $PV \subset \hat{PV}$. Let G' denote the subgroup of G generated by G° and γ_M, δ_M. As above one sees that $G_{M, formal} \subset G'$. The inclusion of the rings of invariants $PV^{G'} \subset \hat{PV}^{G_{M, formal}} = \hat{k}_\infty$ implies that $PV^{G'}$ is a finite field extension of k_∞. This extension is moreover contained in $\hat{k}_\infty$. The conclusion is that $PV^{G'} = k_\infty$ and thus $G' = G$.

3. We have already seen that γ_M and δ_M commute if $\hat{k}_\infty \otimes M$ is mild. The rest of the statement follows from the previous parts of the proposition.   ∎

The following examples show that the map $K \to G/G^o$ can have a non trivial kernel.

Example 10.12 *The order module M with equation $y(z+1) = (1+z^{-1}/n)y(z)$ has the following difference Galois groups:*

   (a) $\mathbf{G}_m$ *over k_∞ (and over $\mathbf{C}(z)$).*

   (b) The cyclic group C_n, generated by γ_M, over the field $\hat{k}_\infty$.

The kernel of $K \to G/G^o$ is generated by γ_M.

Proof: (a) This mild equation is defined over $\mathbf{C}(z)$ and has $\mathbf{G}_m$ as difference Galois group over $\mathbf{C}(z)$ according to example 10.2. Corollary 9.7 implies that its Galois group over k_∞ is also $\mathbf{G}_m$.

   (b) Over $\hat{k}_\infty$ the equation is equivalent to the equation $y(z+1) = (1+z^{-1})^{1/n}$. A solution is $z^{1/n}$ and the difference Galois group is the cyclic group C_n, generated by γ_M.   ∎

Example 10.13 *Let ζ_n be a primitive nth root of unity. The equation $y(z+1) = \zeta_n(1+z^{-2})y(z)$ has difference Galois group $\mathbf{G}_m$ over $\mathbf{C}(z)$ and over k_∞. The difference Galois group over $\hat{k}_\infty$ is cyclic and generated by δ_M. Hence the kernel of $K \to G/G^o$ is generated by δ_M.*

Proof: As in the last example one shows that the difference Galois groups over $\mathbf{C}(z)$ and k_∞ are $\mathbf{G}_m$. Over $\hat{k}_\infty$ the equation is equivalent to $y(z+1) = \zeta_n y(z)$ and has therefore a cyclic difference Galois group generated by δ_M.   ∎

10.4   Mild difference and differential equations

We have seen that very mild differential equations and very mild difference equations over $\hat{k}_\infty$ form equivalent categories. This is no longer true if one replaces $\hat{k}_\infty$ by k_∞. A more precise result is the following.

Proposition 10.14 *The formula $\Phi = exp(\frac{d}{dz})$ induces a tensor functor $\mathcal{F}$ from the category of mild differential modules over k_∞ to the category of the mild difference modules over k_∞. The restriction of this functor to very mild differential equations is fully faithful, but not surjective on (equivalence classes of) objects.*

Proof: Let M be a mild differential module over k_∞. The formula $\Phi = exp(\frac{d}{dz})$ defines an action of Φ on $\hat{k}_\infty \otimes M$. We want to show that Φ is convergent. This is equivalent to showing that $\Phi(M) \subset M$. From the definition of mild it follows that there is a lattice M_0 over $\mathbf{C}\{z^{-1}\}$ in the k_∞-vector space M such that $\frac{d}{dz} M_0 \subset M_0$. One takes a basis $e_1, \dots e_n$ of M_0 over $\mathbf{C}\{z^{-1}\}$. The matrices B and C are the matrices of $\frac{d}{dz}$ and Φ with respect to this basis. The matrix B has coefficients in $\mathbf{C}\{z^{-1}\}$ and the matrix C has coefficients in $\mathbf{C}[[z^{-1}]]$. We have to show that the matrix C is convergent. One can expand $(\frac{d}{dz} + B)^n$ as $\frac{d}{dz}^n + *\frac{d}{dz}^{n-1} + \dots + *\frac{d}{dz} + B(n)$. It is clear from the definition that $C = \sum \frac{1}{n!} B(n)$. We will use this formula to show that C is convergent.

For the $B(n)$ there is a recurrence relation $B(n) = B(n-1)' + BB(n-1)$ and $B(1) = B$. Write $B(n) = \sum_{k \geq 0} B(n)_k z^{-k}$. Let $A \geq 2, c \geq 2$ be constants such that $\|B(1)_k\| \leq cA^k$. By induction one can show that

$$\|B(n)_k\| \leq (n+1)(k+1)^n c^n A^k.$$

Write $C = \sum C_k z^{-k}$ then $\|C_k\| \leq \sum \frac{1}{n!} \|B(n)_k\| \leq e^{2c(k+1)} A^k$. Hence C is convergent.

The functor $\mathcal{F}$ is defined as $\mathcal{F}M$ is equal to M with the action of $\Phi = exp(\frac{d}{dz})$. In order to prove that $\mathcal{F}$ is fully faithful for very mild differential modules, it suffices to show that the map $\mathrm{Hom}(1, M) \to \mathrm{Hom}(\mathcal{F}1, \mathcal{F}M)$ is a bijection if M is a very mild differential module. The left hand side is equal to $\{m \in M | \frac{d}{dz} m = 0\}$ and the right hand side is $\{m \in M | \Phi m = m\}$. For a very mild difference module M over k_∞ the following formula holds in $\dot{k}_\infty \otimes M$:

$$\frac{d}{dz} = \log \Phi = \sum_{n > 0} \frac{(-1)^{n+1}}{n} (\Phi - 1)^n.$$

This shows that the map is bijective. In the next example we will produce a regular difference module over k_∞ which is not isomorphic to the $\mathcal{F}$-image of any very mild differential module over k_∞. ∎

Example 10.15

The regular difference equation of order one, $y(z+1) = c(z)y(z)$ with $c = 1 + *z^{-2} + *z^{-3} + \dots \in k_\infty$ is the image under $\mathcal{F}$ of a regular differential equation $y' = b(z)y$ for some $b = *z^{-2} + *z^{-3} + \dots \in \dot{k}_\infty$. If b were an element of k_∞ then the differential equation $y' = by$ is trivial, i.e. has a solution in k_∞^*. Then $y(z+1) = c(z)y(z)$ has the same non trivial solution. However, we know that a general regular order one equation does not have a solution in k_∞^*. We will produce a more explicit example.

The relation between b and c is given in the proof of 10.14. In this special case this relation is $b(z+1) - b(z) = -\frac{c(z)'}{c(z)}$. The choice $c(z) = e^{-z^{-2}}$ produces the equation $b(z+1) - b(z) = -2z^{-3}$. The "connection matrix" of this equation is, according to 10.4 equal to

$$\mathcal{L}(-2z^{-3}) = \mathcal{L}(\frac{d}{dz}(z^{-2})) = (2\pi i u \frac{d}{du})(\frac{-4\pi^2 u}{(u-1)^2}) \neq 0.$$

This shows that b is divergent.

10.5 Very mild difference modules and multisummability

The difference module M over k_∞ is supposed to be very mild. With the notation of the proof of the last proposition, the action of Φ on M is represented by a matrix $C(z)$ and the action of the corresponding formal differential module is given by a matrix $B(z)$ with coefficients in $\mathbf{C}[[z^{-1}]]$. The matrix B is the unique formal solution of the difference equation

$$B(z+1) - C(z)B(z)C(z)^{-1} = -(\frac{d}{dz}C(z))C(z)^{-1}.$$

This is again a very mild difference equation. The eigenvalues are the $\{g_i g_j^{-1}\}$ where the $\{g_i\}$ are the eigenvalues of the equation $y(z+1) = C(z)^{-1}y(z)$. In order to explain the behavior of the formal matrix B we have to recall some definitions and facts from the theory of *multisummability*. . We refer to [40] for more details. The general definition of multisummability is rather involved. The simple definition, given below, is in fact a theorem.

Let $k > 0$ and let $y = \sum_{n=0}^\infty y_n z^{-n}$ be a formal power series. Let $k > 0$ and let d be a direction at ∞. Then y is called *k-summable in the direction d* if there is a holomorphic function f defined on a (bounded) sector at ∞ with opening $(d - \alpha/2, d + \alpha/2)$ and $\alpha > \frac{\pi}{k}$, and if there is a constant $A > 0$ such that for all $N \geq 1$ and all z in the bounded sector the following inequalities hold

$$|f(z) - \sum_{n=0}^{N-1} y_n z^{-n}| \leq A^N \Gamma(1 + \frac{N}{k})|z|^{-N}.$$

The condition posed on f is much stronger than saying that f has asymptotic expansion y. In fact, the holomorphic function f is unique and is called the *k-sum of y in the direction d*. We note that for $k \leq 1/2$ the f above is in fact a multivalued function defined on a sector with opening greater than 2π. This difficulty can be removed by taking a suitable root of z. We refer to [40] for precise details in this case.

For a sequence of positive numbers $\underline{k} = k_1 < \ldots < k_r$ and a direction d at ∞, the formal power series y is called $\underline{k}$-multisummable in the direction d if y can be written as a sum $y = y_1 + \ldots + y_r$ such that each y_i is k_i-summable in the direction d. The y_i are unique up to holomorphic functions at ∞. This means that we may change each y_i into $y_i + g_i$ with g_i holomorphic at ∞ (and with $\sum g_i = 0$). Let f_i be the k_i-sum in the direction d then $\sum f_i$ is the *multisum in the direction d*. This multisum is unique and lives as a function on a sector with opening $(d - \alpha/2, d + \alpha/2)$ with $\alpha > \frac{\pi}{k_r}$.

According to [16] (see also Theorem 11.1) the formal solution B is $\underline{k}$-multisummable in all but finitely many directions. The sequence $\underline{k} = k_1 < \ldots < 1$ are all the levels present in the eigenvalues $\{g_i g_j^{-1}\}$ of the equation $y(z + 1) = C(z)^{-1} y(z)$.

10.6 Very mild differential modules

(1) Let $(M, \frac{d}{dz})$ be a very mild differential module with corresponding very mild difference module (M, Φ). Then $(M, \frac{d}{dz})$ and (M, Φ) have *"the same asymptotic theory"* for formal solutions. This can be seen as follows.

Let $v \in \hat{k}_\infty \otimes M$ be a solution of $\Phi(v) = v$. Then also $\frac{d}{dz} v = 0$. The asymptotic theory for differential equations asserts that v is multisummable in almost all directions d at ∞. The possible exceptions are the (finitely many) singular directions of $(M, \frac{d}{dz})$. The positive slopes of the differential module M are $\underline{k} = k_1 < \ldots < k_r$ with $k_r < 1$ since the equation is very mild. Then v is $\underline{k}$-multisummable in almost all directions. Let d be a direction for which v is $\underline{k}$-multisummable. Write $v = v_1 + \ldots + v_r$ as above; let f_i be the k_i-sum of v_i in the direction d and put $f = \sum f_i$. Then $\phi(f_i)$ is the k_i-sum of $\Phi(v_i)$ in the direction d. Since $v = \sum \Phi(v_i)$ holds and since this decomposition is unique up to convergent expressions at ∞ we have $\Phi(v_i) = v_i + g_i$ with $g_i \in M$ (i.e. the g_i are convergent) and $\sum g_i = 0$. The unicity of the k_i-sum implies that $\phi(f_i) = f_i + g_i$. Hence $\phi(f) = f$ and so the multisum of v in the direction d is a solution of the difference equation.

(2) Proposition 10.14 implies that the very mild differential module $(M, \frac{d}{dz})$ and the corresponding very mild difference module (M, Φ) have the same Galois group.

(3) We will now discuss the paper [12] of G.D. Birkhoff, which deals precisely with the difference equation $y(z + 1) = C y(z)$ associated with a (very) mild differential equation $y' = By$ over k_∞. Let the invertible holomorphic matrix $Y(z)$, defined on some sector at ∞ satisfy $Y(z)' = BY(z)$. Then Birkhoff states

that also $Y(z + 1) = CY(z)$. This is correct and can be proved in the following way.

One verifies that $(\frac{d}{dz})^n Y(z) = B(n)Y(z)$, where $B(n)$ is the matrix introduced in the proof of Proposition 10.14. As a consequence the infinite expression $\sum \frac{1}{n!}(\frac{d}{dz})^n Y(z)$ converges and has as sum $CY(z)$. The sum is also equal to $Y(z + 1)$.

There is a fundamental matrix $Y_+(z)$ (this means $Y_+(z + 1) = CY_+(z)$), in the notation of [12], having the required asymptotic behavior on a sector $(-\epsilon - \frac{\pi}{2}, +\epsilon + \frac{\pi}{2})$. Then Birkhoff claims that $Y(z) = Y_+(z)D$ for some constant matrix D. In other words, $Y_+(z)$ also satisfies $Y_+(z)' = BY_+(z)$. We will explain why this cannot be correct.

Suppose that this is correct, then the same holds for the other fundamental matrix Y_- defined for a sector $(-\epsilon + \frac{\pi}{2}, +\epsilon + \frac{3\pi}{2})$. The upper and lower connection matrices of the difference equation would then be constant matrices. Those constant matrices are the matrix 1 and the formal monodromy, since those matrices are their values at $u = 0$ and $u = \infty$. The conclusion from Corollary 9.6 is that the difference Galois group of $Y(z+1) = C \, Y(z)$ is equal to the formal difference Galois group. The differential Galois group of $y' = By$ coincides, according to (2), with the difference Galois group of $y(z + 1) = C \, y(z)$ and therefore with the formal difference Galois group of that equation. The last group is also the formal differential Galois group of $y' = By$. The final conclusion is that for any very mild differential equation $y' = By$ at ∞, the differential Galois group coincides with the formal differential Galois group. This is certainly not true. We will give an example.

Consider the order two differential equation $y(z)' = \begin{pmatrix} z^{-1/2} & 0 \\ 0 & -z^{1/2} \end{pmatrix} y(z)$. After a transformation one can change this equation into a form defined over k_∞, namely $y(z)' = \begin{pmatrix} 0 & z^{-1} \\ 1 & -1/2z^{-1} \end{pmatrix} y(z)$. The differential Galois group of this equation is equal to the formal differential Galois group because we have chosen a canonical equation. This group is generated by the two dimensional exponential torus $\{\begin{pmatrix} * & 0 \\ 0 & * \end{pmatrix}\}$ and a formal monodromy matrix $\begin{pmatrix} 0 & 1 \\ 1 & 0 \end{pmatrix}$. A small perturbation of the equation introduces Stokes matrices, which have the form either $\begin{pmatrix} 1 & * \\ 0 & 1 \end{pmatrix}$ or $\begin{pmatrix} 1 & 0 \\ * & 1 \end{pmatrix}$. The theory of J.-P. Ramis and J. Martinet states that there is a small perturbation with a non-trivial Stokes matrix. Moreover, the differential Galois group is generated by the formal differential Galois group and the Stokes matrices. This implies that the differential Galois group of the perturbed equation is $Gl(2, \mathbf{C})$.

Chapter 11

Wild difference equations

11.1   Introduction

The theme of this section is the problem of lifting symbolic solutions of a wild
difference equation to sectors or more general domains in $\mathbf{C}$. The asymptotic behavior of the Gamma function is responsible for complicated analytic problems
which do not occur in the case of a mild difference equation. The aim of "exact
asymptotics" is to find *unique* lifts with additional properties on certain sectors.
Multisummation provides such unique lifts. Recent work of B.L.J. Braaksma
and B.F. Faber [16] proves that, under certain hypotheses, symbolic (or formal)
solutions are multisummable in many directions. We have already used their
results for the analysis of the asymptotic theory of mild equations. As we will
see, one cannot expect multisummability in the general case (at least with the
present definition of multisummability). In fact the rather restrictive hypotheses
of [16] are necessary for multisummability.

This brings us to asking for lifts of symbolic solutions which are not multi-
sums and which are not unique. The important work of G.K. Immink in [31, 32]
will be the basis for our investigations. It is shown in [31, 32] that formal solutions can be lifted to quadrants. Our aim is to find large sectors at ∞ where
formal solutions can be lifted. It is shown that on half planes formal solutions
have asymptotic lifts. One uses a combination of two quadrants to prove the
lifting property for a right half plane, more precisely on a sector of the form
$\{z \in \mathbf{C} \mid -\frac{\pi}{2} < \arg(z) \leq \frac{\pi}{2}\}$. The case of a left half plane is of course similar. Then same method proves the lifting property for the upper and lower half
plane. The proofs use precise information about the asymptotic of the Gamma
function and other functions. Further a method, reminiscent of Cartan's lemma
on analytic functions in several complex variables, is developed. We note that
B.L.J. Braaksma and B.F. Faber are presently working at a direct analytic proof
that half planes have the lifting property.

The results above on the lifting property for half planes lead to a definition of a "connection cocycle". The theorem of Malgrange and Sibuya is the tool for the inverse problem concerning this connection cocycle. This simplifies earlier results in this direction in [32].

11.2   Multisummability of formal solutions

We will first give a formulation of the main result of [16] which fits in our terminology. We want to define the *singular directions* of a difference equation $y(z + 1) = A\ y(z)$ with A an invertible meromorphic matrix at $z = \infty$. For convenience we will include $\pm\frac{\pi}{2}$ in the set of singular directions. The other singular directions are related to the eigenvalues $g \in \mathcal{G}$ of the equation. Let $g = z^\lambda\ c\ exp(\phi(q) - q)\ (1 + z^{-1})^{a_0}$ be an eigenvalue. A direction ϕ (always $-\pi < \phi \le \pi$) is singular for this g if a function representing $e(g)$ has locally maximal descent for $z = re^{i\phi}$ and $r \to +\infty$. For each term separately we will calculate the singular directions. For $c \ne 1$ we can represent $e(c)$ by the function $e^{z(\log(c) + 2\pi i n)}$. If $|c| \ne 1$ then one finds countably many singular directions having as limits the directions $\frac{\pi}{2}, -\frac{\pi}{2}$. If $|c| = 1$ then the only singular directions are $\frac{\pi}{2}, -\frac{\pi}{2}$. The singular directions that we have defined so far are called *level 1*. The term q is a finite sum $\sum_{0 < \mu < 1} a_\mu z^\mu$. We consider each summand separately. The term $e(a_\mu z^\mu)$ can be represented by the finitely many functions $exp(e^{\mu(\log(z) + 2\pi i n)})$. This gives a finite set of singular directions. Those singular directions are called *level μ*. The term $(1 + z^{-1})^{a_0}$ is represented by the functions $e^{a_0(\log(z) + 2\pi i n)}$ and gives no singular directions (c.f. Section 9.2). Let $\underline{k} = k_1 < k_2 < \ldots < 1$ denote the levels that we have found above. The result in [16] can now be formulated as follows.

Theorem 11.1 *Let $\dot{y}$ be a formal solution, i.e. with coefficients in $\mathcal{P}$, of the difference equation $y(z + 1) = A\ y(z) + a$ where A is an invertible meromorphic matrix at ∞ and a is a meromorphic vector at ∞. Let $\lambda_1 < \ldots < \lambda_s$ be the set of rational λ such that the equation has an eigenvalue g of the form $z^\lambda c \ldots$. Let ϕ be a direction which is not singular for the equation. Then y is $\underline{k}$-multisummable in the direction ϕ in the following cases:*

1. *If $\lambda_1 \ge 0$ and $-\frac{\pi}{2} < \phi < \frac{\pi}{2}$.*

2. *If $s = 1$ and $\lambda_1 = 0$.*

3. *If $\lambda_s \le 0$ and either $\frac{\pi}{2} < \phi < \pi$ or $-\pi < \phi < -\frac{\pi}{2}$.*

It is probable that we have introduced too many singular directions and levels. This is not essential for the application Theorem 9.1 of the theorem which is the basis of our analysis of the asymptotics of mild equations in Section 9. In the next example we will show that multisummable of formal solutions is no longer valid in a more general situation.

Example 11.2 *Formal solutions which are not multisummable*

The equation $y(z+1) = zy(z)+1$ has a unique formal solution $\hat{y}_1 \in \mathbb{C}((z^{-1}))$. According to Theorem 11.1, this formal solution is multisummable in the directions $d = e^{i\phi}$ with $-\frac{\pi}{2} < \phi < \frac{\pi}{2}$ and with as multisum the function y_{right} of Section 11.5. According to Section 11.5, y_1 is not multisummable in other directions.

The equation $y(z+1) = z^{-1}y(z)+1$ has a unique formal $\hat{y}_2$. A similar analysis of this equation yields that $\hat{y}_2$ is multisummable precisely in the directions $d = e^{i\phi}$ with $\frac{\pi}{2} < \phi < \frac{3\pi}{2}$.

We combine the two equations into an inhomogeneous matrix equation

$$\begin{pmatrix} y_1 \\ y_2 \end{pmatrix}(z+1) = \begin{pmatrix} z & 0 \\ 0 & z^{-1} \end{pmatrix}\begin{pmatrix} y_1 \\ y_2 \end{pmatrix}(z) + \begin{pmatrix} 1 \\ 1 \end{pmatrix}.$$

There is a unique formal solution $\begin{pmatrix} y_1 \\ y_2 \end{pmatrix}$. This formal solution is multisummable in a direction d if and only if both $\hat{y}_1$ and $\hat{y}_2$ are multisummable in that direction. We conclude that the unique formal solution is in no direction multisummable!

11.3 The Quadrant Theorem

A *quadrant* is a subset of the complex plane of the form

$$v + \{z \in \mathbb{C}\mid z = |z|e^{i\phi}, \; k\frac{\pi}{2} < \phi < (k+1)\frac{\pi}{2}, \; |z| > R\},$$

where $v \in \mathbb{C}$, $k \in \mathbb{Z}$ and $R \geq 0$. We will denote this quadrant by $Q(v, k, R)$. There are essentially four quadrants, they can be shifted over a complex number and a bounded part can be deleted. We note that changing the $<$ signs in the definition of a quadrant into $\leq$ signs is not essential for what follows. A meromorphic function f on $Q(v, k, R)$ is said to have $\sum_{n \geq A} a_n z^{-n/p} \in \mathcal{P}$ as *asymptotic expansion* if there is for every B a constant C such that $|f(z) - \sum_{B > n \geq A} a_n z^{-n/p}| \leq C|z|^{-B/p}$ holds on $Q(v, k, R)$. The following theorem is proved by G.K. Birkhoff and W.J. Trjitzinsky in [14]. This proof contains however a number of inaccuracies and its correctness has been questioned. In [31] and [32] a proof is presented by G.K. Immink. The statement is the following.

Theorem 11.3 *Let the difference equation $y(z+1) = A\ y(z)$, with $A \in Gl(n, \overline{k_\infty})$ and $\overline{k_\infty}$ is the algebraic closure of k_∞, be given. Let A^c be a canonical form for the difference equation. The formal matrix $F \in Gl(n, \mathcal{P})$ with $\hat{F}(z+1)^{-1}AF(z) = A^c$ can be lifted to some invertible meromorphic matrix F on $Q(v, k, R)$ for a suitable $R > 0$, i.e. $F(z+1)^{-1}AF(z) = A^c$ and F has asymptotic expansion $\hat{F}$ on $Q(v, k, R)$.*

We sometimes prefer to work with the following equivalent form of the theorem:

Theorem 11.4 *Let $\hat{y}$ be a formal solution, i.e. with coefficients in $\mathcal{P}$, of the difference equation $y(z + 1) = A\ y(z)$ where $A \in Gl(n, \overline{k_\infty})$ and $\overline{k_\infty}$ is the algebraic closure of k_∞. Let $v \in \mathbf{C}$, $k \in \mathbf{Z}$ also be given. There exists an $R > 0$ and a meromorphic vector y on $Q(v, k, R)$ such that $y(z + 1) = A\ y(z)$ and y has $\hat{y}$ as asymptotic expansion in $Q(v, k, R)$.*

Proof: Let F be as in Theorem 11.3. Using this, our equation $y(z+1) = A\ y(z)$ transforms under F to the equation $v(z + 1) = A^c\ v(z)$ with formal solution $\hat{v}$. Since this equation is in canonical form, v is actually a constant vector. Transforming back with the inverse of F one finds the required y on $Q(v, k, R)$. Theorem 11.3 is in fact equivalent to Theorem 11.4 because F is a formal solution of the difference equation $\hat{F}(z + 1)^{-1}AF(z) = A^c$.   ∎

11.4   On the Gamma function

We will need precise information on the asymptotic behavior of the Gamma function. The classical result is

$$\Gamma(z) \sim e^{-z}e^{(z-1/2)\log(z)}(2\pi)^{1/2}\{1 + \frac{1}{12}z^{-1} + \ldots\},$$

for z with $|\arg(z)| < \frac{\pi}{2} - \epsilon$ and every positive ϵ. The next lemma is concerned with the behavior in an upper strip.

Lemma 11.5 *The Gamma function satisfies in an upper strip, i.e. a set of the form $\{z \in \mathbf{C}|\ a < Re(z) < b,\ Im(z) > c\}$ for real numbers a, b, c and $c > 0$, the inequalities*

$$c_1|z|^{-n} \le |\Gamma(z)|e^{\pi|z|/2} \le c_2|z|^n,$$

for certain positive constants c_1, c_2 and a positive integer n depending on the strip.

Proof: The formula $\Gamma(z+1) = z\Gamma(z)$ shows that we can restrict ourselves to the strip with $a = 0$ and $b = 1$. The product formula reads

$$\Gamma(z)^{-1} = z e^{\gamma z} \prod_{n \geq 1} \{(1 + \frac{z}{n}) e^{-\frac{z}{n}}\}.$$

We will use the notation $f(z) \approx g(z)$ to denote that there are positive constants c_1, c_2 and there is a positive integer n such that $c_1 |z|^{-n} \leq |\frac{f(z)}{g(z)}| \leq c_2 |z|^n$ holds in the region where we the functions f and g are considered. With this notation

$$\Gamma(z)^{-2} \approx \prod_{n \geq 1} |1 + \frac{z}{n}|^2 \approx \prod_{n \geq 1} (1 + \frac{|z|^2}{n^2}) = \frac{\sin(i\pi|z|)}{i\pi|z|}$$

The last expression is equal to $\frac{e^{\pi|z|} - e^{-\pi|z|}}{2\pi|z|} \approx e^{\pi|z|}$. Hence $\Gamma(z) \approx e^{-\pi|z|/2}$. ∎

Remarks 11.6

(1) By conjugation $z \mapsto \bar{z}$ one finds in a "lower strip" the same formula

$$c_1 |z|^{-n} \leq |\Gamma(z)| e^{\pi|z|/2} \leq c_2 |z|^n,$$

for certain positive constants c_1, c_2 and a positive integer n depending on the lower strip.

(2) The behavior of the Gamma function on the left half plane can be found from the behavior on the right half plane by means of the classical formula

$$\Gamma(z) = \frac{-\pi}{z \sin(\pi z) \Gamma(-z)}.$$

In particular the Gamma function has asymptotic expansion 0 on the two quadrants $\{z \in \mathbf{C} | \ Re(z) \leq 0, \ Im(z) \geq c\}$ and $\{z \in \mathbf{C} | \ Re(z) \leq 0, \ Im(z) \leq -c\}$ for any $c > 0$. We will not give the details of the proof.

11.5   An example

An analysis of the equation $y(z+1) = zy(z) + 1$ will be our guide for the study of wild difference equations. It is clear that there is a unique formal solution $\hat{y} \in \mathbf{C}((z^{-1}))$ of the equation. In fact $\hat{y} \in z^{-1}\mathbf{C}[[z^{-1}]]$. The asymptotic lifts on sectors at ∞ and on half planes is what we are studying.

We start with a solution y_{right} of the equation. This solution can be found with the method of Lemma 8.1, namely

$$y_{right} = -\sum_{n \geq 0} z^{-1}(z+1)^{-1} \ldots (z+n)^{-1}.$$

The infinite sum represents a meromorphic function on $\mathbf{C}$ with poles of order one at $0, -1, -2, -3, \ldots$. With the method of Lemma 8.1 one can show that y_{right} has y as asymptotic expansion on the sectors $(-\pi + \epsilon, \pi - \epsilon)$ for every positive ϵ. With more complicated estimates one can show that y_{right} is also an asymptotic lift of $\dot{y}$ on any upper half plane (and any lower half plane) which does not contain the real axis. We want now to find solutions on the "left hand side".

The quadrant theorem asserts that there are asymptotic lifts y_2 and y_3 on the quadrants

$$Q_2 = \{z \in \mathbf{C} | \ Re(z) < 0, \ Im(z) > -1\} \text{ and}$$

$$Q_3 = \{z \in \mathbf{C} | \ Re(z) < 0, \ Im(z) < 1\}.$$

The difference $y_2 - y_3$ is equal to $h(u)\Gamma(z)$ on $Q_2 \cap Q_3$ where h is a meromorphic function of $u = e^{2\pi i z}$, defined for $e^{-2\pi} < |u| < e^{2\pi}$. Furthermore $y_2 - y_3$ has asymptotic expansion 0 on $Q_2 \cap Q_3$. In particular for $z \in Q_2 \cap Q_3$ and $Re(z) << 0$, the function $y_2 - y_3$ has no poles. That implies that h has no poles. Also $h(1) = 0$ since $\Gamma(z)$ has poles in $0, -1, -2, -3, \ldots$. Hence $h = \sum_{n=-\infty}^{\infty} h_n u^n$ and this expression converges on $e^{-2\pi} < |u| < e^{2\pi}$. Put $h^+ := a + \sum_{n>0} h_n u^n$ and $h^- := b + \sum_{n<0} h_n u^n$. The constants a, b are chosen such that $h^+(1) = h^-(1) = 0$. Since $h(1) = 0$ we have $h = h^+ + h^-$.

We claim that $h^+(u)\Gamma(z)$ has asymptotic expansion 0 on Q_2. Write $h^+(u) = (u-1)k(u)$. Then $k(u)$ is bounded for $0 < |u| \leq R$ and any $R < e^{2\pi}$. The function $(u-1)\Gamma(z)$ has no poles. We have to estimate $(u-1)\Gamma(z)$ on Q_2. On the band $\{z \in \mathbf{C} | \ -1 \leq Im(z) \leq 1, \ Re(z) \leq 0\}$ the function $(u-1)\Gamma(z)$ has asymptotic expansion 0 as can be seen from the formula $(u-1)\Gamma(z) = \frac{-\pi(e^{2\pi i z} - 1)}{z \sin(\pi z)\Gamma(-z)}$. On the quadrant $\{z \in \mathbf{C} | \ Re(z) \leq 0, \ Im(z) \geq 1\}$ the function $(u-1)$ is bounded and the Gamma function has asymptotic expansion 0. (See Remarks 11.4, part (2)).

The same arguments show that $h^-(u)\Gamma(z)$ has asymptotic expansion 0 on Q_3. Then $y_2 - h^+(u)\Gamma(z)$ and $y_3 + h^-(u)\Gamma(z)$ are asymptotic lifts of $\dot{y}$ on Q_2 and Q_3. By construction $y_2 - h^+(u)\Gamma(z) = y_3 + h^-(u)\Gamma(z)$ on $Q_2 \cap Q_3$. Thus we have found an asymptotic lift y_{left} of y on the left half plane $Q_2 \cup Q_3$. The function y_{left} is meromorphic on $\mathbf{C}$ since it is a solution of our equation $y(z+1) = zy(z) + 1$. Suppose that f is also an asymptotic lift of $\dot{y}$ on $Q_2 \cup Q_3$. Then the difference $y_{left} - f$ has the form $k(u)\Gamma(z)$ where k is a holomorphic function of u, defined for $u \in \mathbf{C}^*$. Moreover the expansion of $k(u)\Gamma(z)$ in $Q_2 \cup Q_3$ is 0. On an upper strip we use the estimate of Lemma 11.5 for the Gamma function and we conclude that k has no pole at $u = 0$. Similarly k has no pole at $u = \infty$. Thus k is a constant. This constant is 0 because the Gamma function has poles in $0, -1, -2, \ldots$. This shows the unicity of the asymptotic lift y_{left}

on $Q_2 \cup Q_3$. One can find y_{left} explicitly. The function $y_{left} - y_{right}$ has the form $h(u)\Gamma(z)$ with h a holomorphic function of $u \in \mathbf{C}^*$. The function $h(u)\Gamma(z)$ is asymptotically 0 in an upper strip. The behavior of Γ in this strip implies that h has no pole at $u = 0$. Similarly, h has no pole at $u = \infty$. Then h is a constant and $y_{left} = y_{right} + h\Gamma$. The function y_{left} has no pole at $z = 0$. This determines the constant h.

Let f be an asymptotic lift of $\hat{y}$ in a sector at ∞ which has a non empty intersection with the sector $(-\frac{\pi}{2}, \frac{\pi}{2})$, then on the intersection S of the two sectors the function $f - y_{right}$ has asymptotic expansion 0. Write $f - y_{right} = k(u)\Gamma(z)$ where k is meromorphic function of u, which is holomorphic for z with $|z| >> 0$. If S contains the direction $\arg(z) = 0$ then it is rather clear that $k = 0$. If S lies in $(0, \frac{\pi}{2})$ then k is holomorphic for u with $0 < |u| < \delta$ with some positive δ and has a Laurent expansion $k = \sum_{n=-\infty}^{\infty} k_n u^n$. For $t \in S$ such that also $t + 1 \in S$ and any $m \in \mathbf{Z}$ one can form the integral $\int_t^{t+1} k(u)u^{-m}\, dz$. This integral is equal to k_m. The integrand can be estimated on the interval $[t, t+1]$ by $\frac{c|e^{2\pi i t}|}{|\Gamma(t)|}$ for some constant $c > 0$. By shifting $t \in S$ to ∞ one obtains $k_m = 0$. Hence $k = 0$. An analogous reasoning shows that $k = 0$ if the sector S lies in $(-\frac{\pi}{2}, 0)$. We conclude that the assumption on f above implies that $f = y_{right}$.

For an open sector S we say that f is an asymptotic lift of $\hat{y}$ if f satisfies the equation and has $\hat{y}$ as asymptotic expansion on every closed subsector of S. The two sectors $(-\pi, \pi)$ and $(\frac{\pi}{2}, \frac{3\pi}{2})$ (abusing the notation) are then the two maximal open sectors on which $\hat{y}$ has an asymptotic lift. The lifts are y_{right} and y_{left}.

Suppose that $\hat{y}$ is multisummable in the direction $d = e^{i\phi}$ with $\frac{\pi}{2} \le \phi \le \frac{3\pi}{2}$. The multisum y_d in that direction has the correct asymptotic expansion on an open sector S containing $[\phi - \frac{\pi}{2}, \phi + \frac{\pi}{2}]$ because the only level present is 1. This contradicts the statement about the maximal sectors where $\hat{y}$ has an asymptotic lift. We conclude that $\hat{y}$ is not multisummable in the direction d.

The method of this example will be used in the next sections for the construction of asymptotics lifts of formal solutions in the general situation.

11.6 Solutions on a right half plane

The following result is quite close to Theorem 18.13 of [30].

Theorem 11.7 *Let $\hat{y}$ be a formal solution, i.e. with coefficients in $\mathcal{P}$, of the difference equation $y(z + 1) = A\ y(z)$ where $A \in Gl(n, \overline{k_\infty})$ and $\overline{k_\infty}$ is the algebraic closure of k_∞. There exists a meromorphic vector y on a right domain V such that:*

1. $y(z+1) = A\, y(z)$

2. y is holomorphic for $z \in \mathbf{C}$ with $Re(z) \gg 0$.

3. There is a real number b such that for every $\epsilon > 0$ the restriction of y to

$$\{z \in V \mid Re(z) > b \text{ and } -\frac{\pi}{2} + \epsilon \le \arg(z) \le \frac{\pi}{2})\}$$

has $\hat{y}$ as asymptotic expansion.

Proof: As in the proof of Theorem 11.4, it suffices to prove this theorem for a formal solution $\hat{F}$ of the equation $\hat{F}(z+1)^{-1}A\hat{F}(z) = A^c$, where A^c is the canonical form. For the proof of the latter we prefer to work with modules. Let M be the module corresponding to A and M^c the module corresponding to A^c. Then $\hat{F}$ corresponds to an isomorphism $\sigma : \mathcal{P} \otimes M \to \mathcal{P} \otimes M^c$. We take two quadrants Q_1 and Q_4, with $k = 0$ and $k = -1$, which cover a right half plane and have as intersection a band $B := \{z \in \mathbf{C} \mid |Im(z)| < c_1, |z| > c_2\}$ for certain positive numbers c_1, c_2. We may suppose that σ has asymptotic lifts σ_1 and σ_4 on the two quadrants. The map $\sigma_4\sigma_1^{-1}$ is an automorphism of M^c above this band B and is asymptotically the identity. In the next lemma we will show that there are automorphisms τ_1 and τ_4 of M^c such that:

1. τ_1 is defined above Q_1 and is asymptotic to the identity on Q_1.

2. τ_4 is defined above Q_4 and is asymptotic to the identity on
   $\{z \in Q_4 \mid |z| \ge R, \ \arg(z) \in (-\frac{\pi}{2} + \epsilon, 0)\}$ for a fixed R and all $\epsilon > 0$.

3. $\sigma_4\sigma_1^{-1} = \tau_4^{-1}\tau_1$.

Then we change the lifts σ_1 and σ_4 into $\tau_1\sigma_1$ and $\tau_4\sigma_4$. The new lifts coincide on the band and glue therefore to a lift σ_{right} of σ with the asymptotic behavior required in the theorem. Going back to matrices, we have found a meromorphic F on a right half plane with $F(z+1)^{-1}AF(z) = A^c$ and with asymptotic expansion $\hat{F}$ on the regions described in the theorem. The equation satisfied by F shows that F is in fact defined on a right domain. The asymptotic behavior of F implies that F is holomorphic for z with $Re(z) \gg 0$. The asymptotic behavior of F^{-1} shows that F^{-1} has no poles for z with $Re(z) \gg 0$. Hence F is an invertible holomorphic matrix for z with $Re(z) \gg 0$. This shows that our solution y of the original problem is holomorphic for $Re(z) \gg 0$.   ∎

___We suppose now that the difference module M over the algebraic closure $\overline{k}_\infty$ of k_∞ is in canonical form. This means that M is a direct sum of modules $E(g) \otimes M_g$ where g runs in a finite subset of $\mathcal{G}$, where $E(g) = \overline{k}_\infty e_g$ with $\phi(e_g) = ge_g$ and where M_g is a unipotent module, i.e. the operation of Φ has on a special basis the matrix $1 + N\log(1 + z^{-1})$ with N a nilpotent matrix.

Lemma 11.8 *Suppose that M is a difference module in canonical form. Let an automorphism τ of M above the band B be given which is asymptotically the identity. Then there are automorphisms τ_1 and τ_4 of M above the quadrants Q_1 and Q_4 such that:*

1. *τ_1 is asymptotically the identity above Q_1.*

2. *τ_4 is asymptotically the identity for $z \in Q_4$, $|z| \geq R$, with a fixed R, and $\arg(z) \in (-\frac{\pi}{2} + \epsilon, 0)$ for every $\epsilon > 0$.*

3. *$\tau = \tau_4^{-1}\tau_1$.*

Proof: In order to simplify the notations we will suppose that every M_g is one dimensional. In other words, we suppose that M has a basis $e_1, \ldots, e_s$ over the algebraic closure of k_∞ such that $\phi(e_j) = g_j e_j$ where $g_1, \ldots, g_s$ are distinct elements of $\mathcal{G}$. For $g \in \mathcal{G}$ of the form $g = z^\lambda e^{2\pi i (a_0 + i a_1)} exp(\phi(q) - q)(1 + z^{-1})^b$, with $0 \leq a_0 < 1$ and $a_1 \in \mathbf{R}$, we define $e(g)_* = \Gamma(z)^\lambda e^{2\pi i (a_0 + i a_1) z} exp(q) z^b$. For $g = g_j$ all the items in the formula for $e(g)_*$ will be indexed by j. The automorphism τ has the form $\tau(e_k) = \sum_k h_{j,k}(u) e(g_j)_*^{-1} e(g_k)_* e_j$, with each $h_{j,k}(u)$ a holomorphic function of $u = e^{2\pi i z}$ in the domain $\{u \in \mathbf{C} |\ e^{-2\pi c_1} < |u| < e^{2\pi c_1}\}$. The condition that τ is asymptotically the identity translates in to $h_{j,j} - 1$ and $h_{j,k}(u) e(g_j)_*^{-1} e(g_k)_*$, with $j \neq k$, have asymptotic expansion 0 on the band. It follows at once that $h_{j,j} = 1$. For $j \neq k$, one finds inequalities $|h_{j,k}(u)| \leq c_N |e(g_j)_* e(g_k)_*^{-1} z^N|$ for every positive integer N and with a positive constant c_N depending on N. We choose now an ordering of the basis $e_1, \ldots, e_s$ such that for $j < k$ one of the following statements is correct

1. $\lambda_j < \lambda_k$.

2. $\lambda_j = \lambda_k$ and $a_1(j) > a_1(k)$.

3. $\lambda_j = \lambda_k$, $a_1(j) = a_1(k)$ and if there is a μ with $Re(a_\mu) \neq 0$ in the expression $\sum a_\mu z^\mu := q_j - q_k$ then the highest μ with $Re(a_\mu) \neq 0$ satisfies $Re(a_\mu) < 0$.

It is clear that such an ordering exists. The Gamma function has on the band B the asymptotic behavior:

$$\Gamma(z) \sim e^{-z} e^{(z-1/2)\log(z)} (2\pi)^{1/2} \{1 + \frac{1}{12} z^{-1} + \ldots\}.$$

Using this one finds that $h_{j,k} = 0$ for $j < k$. Hence the matrix of τ is an upper triangular matrix with 1's on the diagonal. Our task is to write $\tau = \tau_4^{-1}\tau_1$ with automorphisms τ_4 and τ_1 which are asymptotically the identity above $\{z \in Q_4 | \arg(z) \in (-\frac{\pi}{2} + \epsilon, 0)\}$ (every $\epsilon > 0$) and Q_1. We do this by multiplying the matrix of τ on the right and on the left by a sequence of upper triangular matrices, with 1's on the diagonal, coming from automorphisms which are asymptotically the identity above Q_4 and Q_1. In the first step one wants to kill the entries of the matrix of τ which are on the line above the diagonal. Each of the following steps should remove the entries on a line parallel to the diagonal. One sees that it suffices that to solve the following "*additive problem* ":

Let $j > k$ and let $h(u)e(g_j)_*^{-1}e(g_k)_*$ have asymptotic expansion 0 on the band B. Then h can be written as a sum of h^+ and h^- such that:
  (1) $h^+(u)e(g_j)_*^{-1}e(g_k)_*$ has asymptotic expansion 0 on Q_1.
  (2) $h^-(u)e(g_j)_*^{-1}e(g_k)_*$ has asymptotic expansion 0 on
   $\{z \in Q_4| -\frac{\pi}{2} + \epsilon \le \arg(z) \le 0\}$ for all $\epsilon > 0$.

Put $h = \sum_{n=-\infty}^{\infty} c_n u^n$. The first case to consider is $\lambda_j > \lambda_k$. Then $h^+ := \sum_{n>A} c_n u^n$ and $h^- := \sum_{n \le A} c_n u^n$. We take A sufficiently large. The critical estimate for $h(u)^+ e(g_j)_*^{-1}e(g_k)_*$ is on the strip

$$\{z \in \mathbf{C}| \, a < Re(z) < b, \, Im(z) \ge c\}$$

for real numbers a, b, c with $c > 0$. Using the estimate of Lemma 11.5 for the Gamma function, one finds that on this strip the function $e(g_j)_*^{-1}e(g_k)_*$ behaves like $u^{(\lambda_k - \lambda_j)/4 + a_0(k) - a_0(j)} exp(q_k - q_j)$. The choice $A = -\{(\lambda_k - \lambda_j)/4 + a_0(k) - a_0(j)\}$ guarantees that h^+ satisfies (1). The asymptotic formula for the Gamma function for z with $\arg(z) \in (-\frac{\pi}{2}, \frac{\pi}{2})$ implies that h^- satisfies (2).

The next case to consider is $\lambda_k = \lambda_j$ and $a_1(k) > a_1(j)$. Let $n \in \mathbf{Z}$. The condition that

$$u^n e^{2\pi i(a_0(k) - a_0(j) + i(a_1(k) - a_1(j)))z} exp(q_k - q_j)$$

is asymptotically 0 for $0 \le \arg(z) \le \frac{\pi}{2}$ is equivalent to

$$(n + a_0(k) - a_0(j))\sin(\phi) + (a_1(k) - a_1(j))\cos(\phi) > 0$$

for $0 \le \phi \le \frac{\pi}{2}$. This is the case if $n + a_0(k) - a_0(j) > 0$. The condition that the same expression has asymptotic expansion 0 for $-\frac{\pi}{2} + \epsilon < \arg(z) \le 0$ (every positive ϵ) is true if

$$(n + a_0(k) - a_0(j))\sin(\phi) + (a_1(k) - a_1(j))\cos(\phi) > 0$$

for $-\frac{\pi}{2} < \phi \le 0$. This is the case if $n + a_0(k) - a_0(j) \le 0$. From this it follows that the choice $h^+ = \sum_{n > a_0(j) - a_0(k)} c_n u^n$ and $h^- = \sum_{n \le a_0(j) - a_0(k)} c_n u^n$ has the correct asymptotic properties.

The last case to consider is $\lambda_k = \lambda_j$, $a_1(k) = a_1(j)$ and "if $\sum a_\mu z^\mu := q_k - q_j$ has a term with $Re(a_\mu) \ne 0$ then the largest μ with $Re(a_\mu) \ne 0$ satisfies $Re(a_\mu) <$

0". Let $n \in \mathbf{Z}$. The expression $u^n e(g_j)_*^{-1} e(g_k)_*$ has asymptotic expansion 0 on Q_1 if $n + a_0(k) - a_0(j) > 0$. The same expression $u^n e(g_j)_*^{-1} e(g_k)_*$ has asymptotic expansion 0 for $-\frac{\pi}{2} < \arg(z) \le 0$ if $n + a_0(k) - a_0(j) < 0$ as one can easily verify. Suppose that $n + a_0(k) - a_0(j) = 0$ then either for $0 \le \arg(z) \le \frac{\pi}{2}$ or for $-\frac{\pi}{2} < \arg(z) \le 0$ the expression $u^n e(g_j)_*^{-1} e(g_k)_*$ has asymptotic expansion 0. As in the previous case the choice $h^+ = \sum_{n > a_0(j) - a_0(k)} c_n u^n$ or $h^+ = \sum_{n \ge a_0(j) - a_0(k)} c_n u^n$ and $h^- = h - h^+$ has the correct asymptotic properties. ∎

Remarks 11.9

(1) The methods of the proof of Theorem 11.7 allows the following variation on Theorem 11.7:

There is a solution y of the equation, defined on a right domain, holomorphic for $Re(z) \gg 0$ and with $\hat{y}$ as asymptotic expansion on sets

$$\{z \in \mathbf{C} \mid |z| \ge R, \ -\frac{\pi}{2} \le \arg(z) \le \frac{\pi}{2} - \epsilon\}$$

for a fixed R and all $\epsilon > 0$.

(2) There are two obvious analogues of Theorem 11.7 for left half spaces.

(3) One might think that a somewhat larger sector than a half plane has already the "lifting property". Section 11.5 shows that this is not the case. Indeed, the solution y_{left} does not have $\hat{y}$ as asymptotic expansion on, say, the sector $(\frac{\pi}{2} - \epsilon, \frac{3\pi}{2})$ for $\epsilon > 0$.

(4) Also an extension of Theorem 11.7 to the closed sector $[-\frac{\pi}{2}, \frac{\pi}{2}]$ is in general false. This can easily be seen from the "additive problem" stated in Lemma 11.8.

11.7 Solutions on an upper half plane

The next theorem gives a positive answer to the old question: "Does a formal solution of a difference equation lift to sectors around the directions $\pm\frac{\pi}{2}$?". We will show that a formal solution $\hat{y}$ can be lifted to a solution y on an upper plane $H = \{z \in \mathbf{C} \mid Im(z) > c\}$. More precisely, y has asymptotic expansion $\hat{y}$ on $\{z \in H \mid 0 \le \arg(z) \le \pi - \epsilon\}$ for every positive ϵ. The obvious variant of this: there is a solution y such that y has $\hat{y}$ as asymptotic expansion on $\{z \in H \mid \epsilon \le \arg(z) \le \pi\}$ for every positive ϵ is equally true. The stronger statement that y has the correct asymptotic expansion on the whole of H is probably false. Section 11.5 shows at least that one cannot expect an asymptotic lift on an open sector which is strictly greater than $(0, \pi)$.

Theorem 11.10 *Let $\hat{y}$ be a formal solution, i.e. with coefficients in $\mathcal{P}$, of the difference equation $y(z + 1) = A\ y(z)$ where $A \in Gl(n, \overline{k_\infty})$ and $\overline{k_\infty}$ is the algebraic closure of k_∞. There exists an upper half plane*
$H := \{z \in \mathbf{C}|\ Im(z) > c\}$ *for some $c > 0$ and a meromorphic vector y, defined on a right domain containing H, such that:*

1. *$y(z + 1) = A\ y(z)$.*

2. *y is holomorphic for $Re(z) \gg 0$.*

3. *For every $\epsilon > 0$ the vector y has $\hat{y}$ as asymptotic expansion on $\{z \in H|\ 0 \leq \arg(z) \leq \pi - \epsilon\}$.*

Proof: The proof has the same structure as the proof of Theorem 11.7. We have in fact to prove the analogue of Lemma 11.8. Let a canonical difference module M over the algebraic closure of k_∞ and an automorphism τ of M, defined and asymptotically the identity on an upper strip $\{z \in \mathbf{C}|\ a < Re(z) < b,\ Im(z) > c\}$, be given. We have to show that $\tau = \tau_l \tau_r$, where the automorphism τ_l lives on a left domain and must for every positive ϵ be asymptotically to the identity for $\{z \in H|\ Re(z) < b\}$ and $0 \leq \arg(z) < \pi - \epsilon$. Furthermore the automorphism τ_r must also live above a right domain and must be asymptotically the identity on the set $\{z \in H|\ a < Re(z)\}$.

For notational convenience we suppose that the canonical module M has a basis $e_1, \ldots, e_s$ over the algebraic closure of k_∞ such that $\phi(e_j) = g_j e_j$, where $g_1, \ldots, g_s$ are distinct elements of $\mathcal{G}$. For an element

$$g = z^\lambda e^{2\pi i(a_0 + ia_1)} exp(\phi(q) - q)(1 + z^{-1})^b \in \mathcal{G},$$

with $a_0, a_1 \in \mathbf{R}$ and $|a_0| < 1$, we write $e(g)_* = \Gamma(z)^\lambda e^{2\pi i(a_0 + ia_1)z} exp(q) z^b$.

As in Section 11.5, one can show that if $(\sum_{n=-\infty}^{\infty} h_n u^n) e(g)_*$ is asymptotically 0 on the strip then each term $h_n u^n e(g)_*$ has also this property. Our first concern is therefore to find out when $u^n e(g)_*$ is asymptotically 0 in the strip. Using the estimate of Lemma 11.5 for the Gamma function in the strip, one finds that the integers n are given by:

(a) $\frac{\lambda}{4} + a_0 + n > 0$ and

(b) $\frac{\lambda}{4} + a_0 + n = 0$ if q has the special property (*). Property (*) is the following. Write $q = \sum a_\mu z^\mu$. If there is μ with $Re(a_\mu i^\mu) \neq 0$ then the largest μ with $Re(a_\mu i^\mu) \neq 0$ satisfies $Re(a_\mu i^\mu) < 0$.

In the next calculation we want to know whether an expression $u^n e(g)_*$ which is asymptotically 0 on the strip is also asymptotically 0 on $\{z \in \mathbf{C}|\ Im(z) > 0,\ Re(z) < b,\ 0 \leq \arg(z) < \pi - \epsilon\}$ for every positive ϵ. The last property will be called *left-flat*. The results are the following:

1. For $\lambda < 0$ the expression $u^n e(g)_*$ is *not* left-flat.

2. For $\lambda > 0$ the expression $u^n e(g)_*$ is left-flat if $\frac{\lambda}{4} + a_0 + n > 0$. The expression is also left-flat if $\frac{\lambda}{4} + a_0 + n = 0$ and q satisfies (*).

3. For $\lambda = 0$ and $a_1 < 0$ the expression is left-flat if $a_0 + n > 0$ and also for $a_0 + n = 0$ (in that case $a_0 = n = 0$) if q satisfies (*).

4. For $\lambda = 0$ and $a_1 > 0$ the expression is *not* left-flat.

5. For $\lambda = 0$ and $a_1 = 0$ the expression is left-flat if $a_0 + n > 0$. If $a_0 + n = 0$, and so $a_0 = n = 0$, the expression if left-flat if q has the property $Re(q(z)) < 0$ for z with $Im(z) > 0$ and $\frac{\pi}{2} \leq \arg(z) < \pi$. This property of q will be denoted by (*l).

We will call a function *right-flat* if the function has asymptotic expansion 0 on the set $\{z \in \mathbb{C} | \; Im(z) > 0, \; Re(z) > b\}$. Let the expression $u^n e(g)_*$ have asymptotic expansion 0 on the strip. The question whether $u^n e(g)_*$ is right-flat has the following answer:

1. For $\lambda > 0$ the expression $u^n e(g)_*$ is *not* right-flat.

2. For $\lambda < 0$ the expression $u^n e(g)_*$ is right-flat if $\frac{\lambda}{4} + a_0 + n > 0$. The expression is also right-flat if $\frac{\lambda}{4} + a_0 + n = 0$ and q satisfies (*).

3. For $\lambda = 0$ and $a_1 > 0$ the expression is right-flat if $a_0 + n > 0$ and also for $a_0 + n = 0$ (in that case $a_0 = n = 0$) if q satisfies (*).

4. For $\lambda = 0$ and $a_1 < 0$ the expression is *not* right-flat.

5. For $\lambda = 0$ and $a_1 = 0$ the expression is right-flat if $a_0 + n > 0$. If $a_0 + n$, and so $a_0 = n = 0$, the expression if right-flat if q has the property $Re(q(z)) < 0$ for z with $Im(z) > 0$ and $0 \leq \arg(z) \leq \frac{\pi}{2}$. This property of q will be denoted by (*r).

One can easily analyse the properties (*),(*l) and (*r). If q satisfies (*) then q satisfies (*l) or (*r) (or both). This shows that if $u^n e(g)_*$ asymptotically 0 on the strip then $u^n e(g)_*$ is left-flat or right-flat (or both). For the $g_1, \ldots, g_s \in \mathcal{G}$ above we choose the $e(g_j)_*$ as above but now with the condition $0 \leq a_0(j) < 1$. The ordering of the basis $e_1, \ldots, e_s$ is chosen such that $j \leq k$ implies that $e(g_j)_*^{-1} e(g_k)_*$ has the property:

If $u^n e(g_j)_*^{-1} e(g_k)_*$ is asymptotically 0 on the strip then $u^n e(g_j)_*^{-1} e(g_k)_*$ is left-flat.

From the results above one deduces that such an ordering exists. The given automorphism τ of M above the strip commutes with ϕ and has therefore the matrix

$$\tau(e_k) = \sum_j h_{j,k}(u)e(g_j)_*^{-1}e(g_k)_* e_j \ ,$$

where the $h_{j,k}$ are holomorphic functions of u for $0 < |u| < \delta$ for a certain $\delta > 0$. For $j \neq k$ one has that $h_{j,k}(u)e(g_j)_*^{-1}e(g_k)_*$ is asymptotically 0 on the strip. Also $h_{j,j} - 1$ is asymptotically 0 on the strip. It follows that the $h_{j,k}$ have at most poles at $u = 0$. In other words the functions $h_{j,k}$ belong to the field $\mathbf{C}(\{u\})$. The matrix of τ is asymptotically the identity on the strip. From this it follows that for any r with $1 \leq r \leq s$ the determinant of the matrix $(h_{j,k}(u)e(g_j)_*^{-1}e(g_k)_*)_{1 \leq j,k \leq r}$ is non zero. The matrix $(h_{j,k})$ inherits this property, i.e. for every r with $1 \leq r \leq s$ the determinant $\det(h_{j,k})_{1 \leq j,k \leq r}$ is not zero. With induction on the size of the matrix $(h_{j,k})$ one can show that there are unique invertible matrices $(a_{j,k})$ and $(b_{j,k})$ with coefficients in the field $\mathbf{C}(\{u\})$ such that

1. $(a_{j,k})(b_{j,k}) = (h_{j,k})$.

2. $a_{j,k} = 0$ if $j > k$.

3. $b_{j,k} = 0$ if $j < k$.

4. $b_{j,j} = 1$ for all j.

Define τ_l and τ_r by the matrices

$$\tau_l(e_k) = \sum_j a_{j,k}(u)e(g_j)_*^{-1}e(g_k)_* e_j \text{ and}$$

$$\tau_r(e_k) = \sum_j b_{j,k}(u)e(g_j)_*^{-1}e(g_k)_* e_j$$

Clearly $\tau_l \tau_r = \tau$. The orders of the functions $h_{j,k}$ at the point $u = 0$ are known. From this one can calculate the orders of the functions $a_{j,k}$ and $b_{j,k}$ at $u = 0$ and verify that τ_l and τ_r have the required asymptotic behavior.   ∎

Remark 11.11 *Theorem 11.10 has two analogues for a lower half plane.*

11.8   Analytic equivalence classes of difference equations

In this subsection we study the set of difference equations which are formally equivalent to a fixed difference equation $y(z+1) = S\, y(z)$. The invertible matrix is supposed to have coefficients in the algebraic closure of k_∞. We consider the pairs (A, F) with the properties:

- A is an invertible matrix with coefficients in the algebraic closure of k_∞.

- $\hat{F}$ is an invertible matrix with coefficients in the field of Puiseux series $\mathcal{P}$.

- $\hat{F}(z+1)^{-1} A \hat{F}(z) = S$.

For such a pair, the equation $y(z+1) = A\, y(z)$ is formally equivalent to the equation $y(z+1) = S\, y(z)$, and $\hat{F}$ is a choice for the formal equivalence.
On this set of pairs we introduce also an equivalence relation. The pairs $(A_i, \hat{F}_i)$, $i = 1, 2$ are called equivalent if there is an invertible meromorphic matrix C, defined over the algebraic closure of k_∞, such that $C(z+1)^{-1} A_1 C(z) = A_2$ and $\hat{F}_1 = C\hat{F}_2$. The set of equivalence classes will be denoted by $\mathrm{Eq}(S)$.

The idea for the study of $\mathrm{Eq}(S)$ is to lift the formal $\hat{F}$ on sectors at ∞ and to compare the various lifts on the intersection of the sectors. This comparison leads to a cocycle for a certain sheaf of groups $\mathrm{Aut}(S)^0$ on the circle S^1 of the directions at ∞. This sheaf is defined as follows:

Let (a, b) be an open interval of S^1, then $\mathrm{Aut}(S)^0(a, b)$ consists of the invertible meromorphic matrices T, defined on a sector at ∞ corresponding to (a, b), such that T has asymptotic expansion 1 for every closed subsector and such that $T(z+1)^{-1} S T(z) = S$.

For the convenience of the reader we recall the definitions for the first cohomology *set* of a sheaf of *non abelian* groups. Let $S_1, \ldots, S_n$ be a covering of the circle S^1 by open intervals. A *cocycle* with respect to this covering is a family of elements $\{\xi_{i,j}\}$, such that

- Each $\xi_{i,j}$ is a section of the sheaf $\mathrm{Aut}(S)^0$ above $S_i \cap S_j$.

- $\xi_{i,j} = \xi_{j,i}^{-1}$.

- On $S_i \cap S_j \cap S_k$ the equality $\xi_{i,j}\xi_{j,k}\xi_{k,i} = 1$ holds.

The *trivial cocycle* is the cocycle with all $\xi_{i,j} = 1$. Two cocycles $\{\xi_{i,j}\}$ and $\{\eta_{i,j}\}$ are called equivalent if there are sections U_i of our sheaf above S_i such that $\eta_{i,j} = U_i^{-1}\xi_{i,j}U_j$ for all i, j.

The set of all equivalence classes of cocycles with respect to the given covering of S^1 is the first Čech cohomology set of the sheaf with respect to this covering. The direct limit, over all coverings of the circle, of those Čech cohomology sets, is the first Čech cohomology set of the sheaf on S^1. This set will be denoted by $H^1(S)$. We note that the map from the cohomology set with respect to a fixed covering to $H^1(S)$ is injective.
It suffices to consider coverings $S_1, \ldots, S_n$ by open intervals of the circle, such

that there are no triple intersections. One may then suppose that S_i has only a non trivial intersection with S_{i-1} and S_{i+1}, where we have introduced the convenient "cyclic" notation $S_i = S_{i+kn}$ for any $k \in \mathbf{Z}$. A cocycle can now be represented by $(\xi_{1,2}, \xi_{2,3}, \ldots, \xi_{n-1,n}, \xi_{n,1})$. There are no longer conditions on the elements $\xi_{i,i+1}$.

For the special situation that concerns us, it suffices to consider the covering $\{S_1, S_2, S_3, S_4\}$ given by the open intervals $(-\frac{\pi}{2}, \frac{\pi}{2}), (0, \pi), (\frac{\pi}{2}, \frac{3\pi}{2})$ and $(-\pi, 0)$. For each S_i and each open covering of S_i, one can show that any cocycle is equivalent with a trivial one. This is a consequence of the lifting property for the sectors S_i.

It follows that every element of $H^1(S)$ can be represented by a cocycle with respect to the covering $\{S_1, S_2, S_3, S_4\}$. In other words, we find the following description of $H^1(S)$:

> $H^1(S)$ is the set of equivalence classes of 4-tuples
> $(T_{1,2}, T_{2,3}, T_{3,4}, T_{4,1})$, where $T_{i,i+1}$ is a section of the sheaf $\mathrm{Aut}(S)^0$.
> Two 4-tuples $\{A_{i,i+1}\}$ are $\{B_{i,i+1}\}$ are equivalent if there are sections
> U_i of the sheaf above S_i such that $B_{i,i+1} = U_i^{-1} A_{i,i+1} U_{i+1}$ for all i.

Theorem 11.12 *There is a natural bijection $Eq(S) \to H^1(S)$.*

Proof: Let a pair $(A, \hat{F})$ be given. We know that there are asymptotic lifts F_i of $\hat{F}$ on the sectors S_i for $1 = 1, \ldots, 4$. We associate to the pair $(A, \hat{F})$ the cocycle $(F_1^{-1} F_2, F_2^{-1} F_3, F_3^{-1} F_4 . F_4^{-1} F_1)$. The class of this cocycle does not depend on the choice of the F_i. Equivalent pairs produce equivalent cocycles. Hence there is a well defined map $Eq(S) \to H^1(S)$. The injectivity of this map is an easy exercise. The surjectivity however is not at all clear. We will use the theorem of Malgrange and Sibuya to prove this. Let a cocycle $(T_{1,2}, T_{2,3}, T_{3,4}, T_{4,1})$ be given. This can also be seen as a cocycle for the sheaf H introduced in Section 8.4. Hence there is a $\hat{F} \in GL(n, \mathbf{C}((z^{-1})))$ and there are lifts F_i of $\hat{F}$ such that $F_i^{-1} F_{i+1} = T_{i,i+1}$ for all i. The matrices $A_i = F_i(z+1) S F_i(z)^{-1}$ are invertible meromorphic on the sectors S_i. On the intersections $S_i \cap S_{i+1}$ the equality $A_i = A_{i+1}$ holds (by construction). The A_i glue to an invertible matrix with coefficients in the algebraic closure of k_∞. The pair $(A, \hat{F})$ has obviously the same image in $H^1(S)$ as the cocycle $(T_{1,2}, T_{2,3}, T_{3,4}, T_{4,1})$.   ∎

Corollary 11.13 *Suppose that the coefficients of S are in $\mathbf{C}(z)$. Let the cocycle $(T_{1,2}, T_{2,3}, T_{3,4}, T_{4,1})$ satisfy:*

1. *$T_{1,2}$ and $T_{2,3}$ are meromorphic on an upper half plane.*

2. *$T_{3,4}$ and $T_{4,1}$ are meromorphic on a lower half plane.*

3. *$T_{1,2} T_{2,3}$ and $T_{3,4} T_{4,1}$ are meromorphic on all of $\mathbf{C}$ and their product is 1.*

Then there is a pair $(A, \hat{F})$ with the same image in $H^1(S)$ as the cocycle, such that A has coefficients in $\mathbf{C}(z)$.

Proof: With the notation of the proof of the last theorem one has that $T_{i,i+1} = F_i^{-1}F_{i+1}$. The matrices F_1, F_2, F_3, F_4 are defined on a right domain, an upper half plane, a left domain and a lower half plane. The F_i are the asymptotic lifts on the sectors S_i of some $\hat{F} \in Gl(n, \mathbf{C}((z^{-1})))$. The expression $F_1^{-1}F_3$ is defined on an upper half plane and on a lower half plane, and is equal to $T_{1,2}T_{2,3}$ and $(T_{3,4}T_{4,1})^{-1}$ on those half planes. The third condition states that there is an invertible meromorphic matrix defined on all of $\mathbf{C}$ which has $F^{-1}F_3$ as restriction on an upper half plane and a lower half plane. Therefore F_1 and F_3 are meromorphic outside a compact subset of $\mathbf{C}$.

The matrix F_1 is invertible meromorphic on $r < |z| < \infty$. We consider the open covering $\{\mathbf{C}, \{z|\ r < |z| \le \infty\}\}$ of the complex projective line $\mathbf{P}^1(\mathbf{C})$. Using Corollary 12.8, one concludes that $F_1 = CD$ with C and D invertible meromorphic matrices on $\{z|\ r < |z| \le \infty\}$ and $\mathbf{C}$ respectively. In particular, there is an invertible meromorphic matrix B at ∞ such that BF_1 is meromorphic on all of $\mathbf{C}$.

The $\{BF_i\}$ are asymptotic lifts on the sectors S_i of $B\hat{F}$. The cocycle associated with the $\{BF_i\}$ is again $(T_{1,2}, T_{2,3}, T_{3,4}, T_{4,1})$. Define the matrices A_i by $A_i := (BF_i)(z+1)S(BF_i)(z)^{-1}$. The matrices A_i and A_{i+1} coincide on the intersections $S_i \cap S_{i+1}$. Hence, there is an invertible meromorphic matrix A at ∞ which has the A_i as restrictions. Since BF_1 is meromorphic on $\mathbf{C}$, the matrix A is also meromorphic on $\mathbf{C}$. Thus the coefficients of A are in $\mathbf{C}(z)$. ∎

Remark 11.14

We will show that the conditions in Corollary 11.13 are also *necessary* for the existence of a pair (A, F) with the properties stated there.

Let the pair $(A, \hat{F})$ be given. The lifts F_i of F have the properties: F_1 is meromorphic on all of $\mathbf{C}$, F_2 is meromorphic on an upper half plane, F_3 is meromorphic on all of $\mathbf{C}$ and F_4 is meromorphic on some lower half plane. Hence $F_1^{-1}F_2$ and $F_2^{-1}F_3$ are meromorphic on an upper half plane. Furthermore $F_3^{-1}F_4$ and $F_4^{-1}F_1$ are meromorphic on a lower half plane. The products $(F_1^{-1}F_2)(F_2^{-1}F_3)$ and $(F_3^{-1}F_4)(F_4^{-1}F_1)$ are equal to $F_1^{-1}F_3$ and $F_3^{-1}F_1$ and so the third condition in Corollary 11.13 is also necessary.

A Problem.

Let $y(z + 1) = A\ y(z)$ be a difference equation, say over k_∞, which has the canonical form $y(z + 1) = A^c\ y(z)$ over k_∞. Suppose that an invertible $\hat{F} \in Gl(n, \hat{k}_\infty)$ is given with $\hat{F}(z+1)^{-1}A\hat{F}(z) = A^c$. If the equation $y(z+1) = A^c\ y(z)$ is semi-regular, or more generally satisfies the conditions of Lemma 8.1, then there are unique asymptotic lifts of F on large sectors. With those lifts and the knowledge of the difference Galois group of the equation $y(z + 1) = A^c\ y(z)$, which is the formal difference Galois group of the equation $y(z + 1) = A\,y(z)$, one can determine the difference Galois group of the equation $y(z + 1) = A\,y(z)$. For mild equations the same method works if one uses multisummation. In both cases, one has produced a canonical choice, i.e. functorial and commuting with tensor products, of a cocycle representing the image of the pair $(A, \dot{F})$ in $H^1(A^c)$.

In the general case it seems that there is no canonical choice. Is there another method to calculate from the canonical equation $y(z + 1) = A^c\,y(z)$ and the image of the pair (A, F) in $H^1(A^c)$ the difference Galois group of the equation $y(z + 1) = A\,y(z)$?

11.9   An example

The following example illustrates methods and results from the last chapters. Consider the mild difference equation

$$y(z + 1) = cy(z) + z^{-1} \text{ with } c \in \mathbf{R} \text{ and } c > 1.$$

There is a unique formal solution $\hat{y}$. It has the form $\sum_{n=1}^{\infty} a_n z^{-n}$. Our aim is to find and compare solutions on sectors which are lifts of $\hat{y}$.

With the elementary method of the proof of Lemma 8.1 one can make a guess for an asymptotic lift v_1 of $\hat{y}$, namely

$$v_1(z) := -\sum_{n=0}^{\infty} \frac{c^{-n-1}}{z + n}.$$

It is easily seen that this sum converges locally uniformly on $\mathbf{C}$ and is a meromorphic function on $\mathbf{C}$. The function has simple poles at the points $0, -1, -2, -3, \ldots$. Further $v_1(z + 1) = cv_1(z) + z^{-1}$. On the sector $(-\pi, \pi)$ the function v_1 has asymptotic expansion $\hat{y}$. We recall the precise meaning of this:

For every $\epsilon > 0$ and every $N > 0$ there are positive constants C, R, depending on ϵ, such that for z with $|z| \geq R$ and $\arg(z) \in (-\pi + \epsilon, \pi - \epsilon)$ the inequality

$$\left| v_1(z) - \sum_{n=1}^{N-1} a_n z^{-n} \right| \leq C|z|^{-N} \text{ holds.}$$

One can verify this by using the expansion $\frac{c^{-n-1}}{z+n} = z^{-1} \sum_{k \geq 0} c^{-n-1}(-n)^k z^{-k}$. The function v_1 is not meromorphic at ∞ since it has poles at the points $0, -1, -2, -3, \ldots$. Therefore $\hat{y}$ is a *divergent* power series and has explicitly the form

$$\hat{y} = \sum_{n=1}^{\infty} (\sum_{k \geq 0} \frac{k^{n-1}}{c^{k+1}})(-z)^{-n}.$$

We are now looking for asymptotic lifts of $\hat{y}$ in a sector S at ∞ which contains the direction π. Suppose that there is an asymptotic lift v on a sector S of the form $(-\epsilon + \frac{\pi}{2}, \frac{3\pi}{2} + \epsilon)$, with some abuse of notation and with $\epsilon > 0$ and small. (We note that multisummation or the methods of this chapter, guarantees an asymptotic lift on a smaller sector). The equation for v shows that v is meromorphic on $\mathbf{C}$ and that the poles of v are simple and form a subset of $\mathbf{Z}$. Then $v - v_1 = e^{z \log(c)} h(u)$, where $\log(c)$ is a positive real number, $u := e^{2\pi i z}$ and h is a meromorphic function of $u \in \mathbf{C}^*$. On the set $\mathbf{C}^*$ the function h has only a simple pole at $u = 1$. Thus $h(u) = \frac{cst}{u-1} + \sum_{n=-\infty}^{\infty} h_n u^n$, with cst some constant and the infinite sum converging for $u \in \mathbf{C}^*$. We are going to use that $v - v_1$ has asymptotic expansion 0 for every direction ϕ with ϕ in the union of the two sectors $(\frac{\pi}{2} - \epsilon, \frac{\pi}{2})$ and $(-\frac{\pi}{2}, -\frac{\pi}{2} + \epsilon)$. For the first sector we write

$$\sum_{n=-\infty}^{\infty} k_n u^n := -cst \sum_{n \geq 0} u^n + \sum_{n=-\infty}^{\infty} h_n u^n$$

for the Laurent expansion of h at $u = 0$. The asymptotic expansion of $\sum k_n u^n e^{z \log(c)}$ is 0 in the first sector. This implies that every term $k_n u^n e^{z \log(c)}$ has asymptotic expansion 0 in the first sector. Take ϕ in the first sector and write $z = re^{i\phi}$ with $r > 0$. If $k_n \neq 0$ then the limit for $r \to \infty$ of the expression $|u^n e^{z \log(c)}| = exp(r(-2\pi n \sin(\phi) + \cos(\phi) \log(c)))$ is zero. This is equivalent with $n \geq \frac{\log(c)}{2\pi} \frac{\cos(\epsilon)}{\sin(\epsilon)}$. For ϵ small enough this is equivalent with $n > 0$. We conclude that $k_n = 0$ for $n \leq 0$. Thus $h_0 = cst$ and $h_n = 0$ for $n < 0$.

The same calculations, but now with ϕ in the sector $(-\frac{\pi}{2}, -\frac{\pi}{2} + \epsilon)$ and the Laurent expansion

$$\sum_{n=-\infty}^{\infty} l_n u^n := cst \sum_{n \geq 1} u^{-n} + \sum_{n=-\infty}^{\infty} h_n u^n \text{ of } h \text{ at } \infty,$$

lead to $l_n = 0$ for $n \geq 0$. In other words $h_n = 0$ for $n \geq 0$. Combining the results above one concludes that $v = v_1$. However v_1 does not have $\hat{y}$ as asymptotic expansion in the direction π because of the presence of the poles at $0, -1, -2, -3, \ldots$. This contradiction shows that:

There is no asymptotic lift of $\hat{y}$ on a sector of the form $(-\epsilon + \frac{\pi}{2}, \frac{3\pi}{2} + \epsilon)$. Even stronger, there is no asymptotic lift on the closed sector $[\frac{\pi}{2}, \frac{3\pi}{2}]$.

We make a small variation on the example above. Consider the equation

$$
\begin{pmatrix} y_1 \\ y_2 \end{pmatrix} (z+1) = \begin{pmatrix} c & 0 \\ 0 & c^{-1} \end{pmatrix} \begin{pmatrix} y_1 \\ y_2 \end{pmatrix} (z) + \begin{pmatrix} z^{-1} \\ z^{-1} \end{pmatrix}
$$

This matrix equation has a unique formal solution. From the calculations above one concludes that this formal solution has no asymptotic lift on either one of the two closed sectors $[-\frac{\pi}{2}, \frac{\pi}{2}]$ and $[\frac{\pi}{2}, \frac{3\pi}{2}]$.

It is interesting to compare this with the statements of G.D. Birkhoff concerning asymptotic lifts, as given in [9] and [12]. In [12] it is stated that a difference equation in matrix form (a mild one over $\mathbf{C}(z)$ in our terminology) has two principal matrix solutions, Y_+ and Y_-, having asymptotic expansions on the sectors $(-\frac{\pi}{2} - \epsilon, \frac{\pi}{2} + \epsilon)$ and $(\frac{\pi}{2} - \epsilon, \frac{3\pi}{2} + \epsilon)$. This is apparently not the case for our example above.

The two principal solution matrices Y_+, Y_- are also present in [9]. It seems that Birkhoff only claims that the two matrices have the correct asymptotic behavior on a right half plane and on a left half plane. The two principal solutions have an additional condition on the location of the poles and are then unique.

The example above forces us to the interpretation of "half plane" as "open half plane". This has the disadvantage that Birkhoff's "connection matrix" $Y_-^{-1} Y_+$ does not have an asymptotic expansion 1 on suitable subsets of $\mathbf{C}$. In particular one cannot conclude that this connection matrix has coefficients in $\mathbf{C}(u)$ with $u = e^{2\pi i z}$, if the original equation has coefficients in $\mathbf{C}(z)$.

However, Birkhoff's results are valid for a more restricted class of difference equations. Moreover, the condition on the location of the poles of the two principal matrix solutions seems to play an important role in his work.

We continue the analysis of the example $y(z+1) = cy(z) + z^{-1}$ using the *method of multisummation* (see Section 10.5 and Theorem 11.1). The singular directions for the multisummation are $\pm\frac{\pi}{2}$ and the directions ϕ_n, where $e^{z(\log(c)+2\pi i n)}$ (with $n \in \mathbf{Z}$) has locally maximal decrease for $z = re^{i\phi_n}$ and $r \to +\infty$.

This condition on ϕ_n is equivalent with: the function $\log(c)\cos(\phi) - 2\pi n \sin(\phi)$ has a negative minimum at ϕ_n. Clearly $\phi_0 = \pi$. In general, ϕ_n is the solution in the interval $(\frac{\pi}{2}, \frac{3\pi}{2})$ of the equation $\mathrm{arctg}(\phi_n) = -2\pi n$.
All the singular directions have level 1. The limit of ϕ_n for $n \to +\infty$ is $\frac{\pi}{2}$ and the other limit is $\frac{3\pi}{2}$.

For a direction d which is not singular, the multisum of $\hat{y}$ (which is here 1-summation or the Borel-1 sum) in the direction d exists and is an asymptotic lift of $\hat{y}$ on a sector $(d - \frac{\pi}{2} - \delta, d + \frac{\pi}{2} + \delta)$ for some positive δ. The multisum for the directions d in an interval of directions (a, b), which does not contain a singular direction, is independent of d.

We apply this to the direction $d = 0$ and we find that the multisum in the direction 0 has the correct asymptotic behavior on the sector $(-\pi, \pi)$. Of course this multisum is v_1.

The interval (ϕ_1, π) does not contain a singular direction. The corresponding multisum, say v_2 has the correct asymptotic behavior on the sector $(\phi_1 - \frac{\pi}{2}, \frac{3\pi}{2})$. Another interesting multisum v_3 of $\hat{y}$, with respect to the interval of directions (π, ϕ_{-1}), has the correct asymptotic behavior in the sector $(\frac{\pi}{2}, \phi_{-1} + \frac{\pi}{2})$.

More generally, we can take two singular directions a, b, such that the interval (a, b) does not contain any singular direction. The multisum in the direction d, with $d \in (a, b)$, does not depend on d. This multisum is an asymptotic lift of $\hat{y}$ and has the required asymptotic expansion on the sector $(a - \frac{\pi}{2}, b + \frac{\pi}{2})$. In our case this is a maximal sector for which there exists an asymptotic lift of $\hat{y}$.

We now propose to use the *method of this chapter* for the study of the asymptotic lifts. The results imply that for every formal solution of a difference equation and any direction d there is an asymptotic lift on a small sector $(d - \delta, d + \delta)$, with $\delta > 0$, around d. In order to find out whether the formal solution has also an asymptotic lift on a given (maybe large) sector, one is led to an investigation of the first cohomology group of the sheaf of solutions of the homogeneous equation which have asymptotic expansion zero. We will explain this method in detail for our example.

Let S^1 denote the circle of the directions at ∞. On S^1 one considers the sheaf of complex vector spaces $\mathcal{S}$, defined as $\mathcal{S}(a, b)$ is the set of the solutions of the equation $y(z + 1) = cy(z)$ which have asymptotic expansion zero on the sector (a, b). Consider the problem:

Does $\hat{y}$ have an asymptotic lift on the sector (a, b)?

There is an open covering $\{S_i\}$ of (a, b) by intervals S_i, for which there are asymptotic lifts y_i. The set $\{y_i - y_j\}$ is a 1-cocycle for the sheaf $\mathcal{S}$ with respect to the covering $\{S_i\}$. If the image of this 1-cocycle is 0 in $H^1((a, b), \mathcal{S})$ then one concludes that there is an asymptotic lift of $\hat{y}$ on the sector (a, b).

In the following we investigate the sheaf $\mathcal{S}$ and the sectors (a, b) for which the cohomology groups $H^1((a, b), \mathcal{S})$ are 0.

Every element of $\mathcal{S}(a, b)$ has the form $e^{z \log(c)} h(u)$, where h is a meromorphic function of $u = e^{2\pi i z}$. If (a, b) is contained in $(0, \pi)$ then h is holomorphic on $0 < |u| < \delta$ for some $\delta > 0$. The function h has a Laurent expansion $\sum_{n=-\infty}^{\infty} h_n u^n$ at $u = 0$. The interval $(a, b) \subset (0, \pi)$ determines which terms are present in this Laurent series. More precisely, the term u^n can have a non zero coefficient if and only if the expression

$$-2\pi n \sin(\phi) + \cos(\phi) \log(c)$$

is negative for every $\phi \in (a, b)$. Using this one can show that

- $\mathcal{S}(a, b) = \mathcal{S}(a, \pi)$ for $0 \le a < b \le \pi$.

- $\mathcal{S}(0, \pi) = 0$.

- $\mathcal{S}(d, \pi) \subset \mathcal{S}(e, \pi)$ for $0 \le d < e < \pi$.

The conclusion is that the sheaf $\mathcal{S}$ has trivial cohomology on any subset of the sector $(0, \pi)$. For similar reasons, the sheaf $\mathcal{S}$ has trivial cohomology on any subset of the sector $(-\pi, 0)$.

It is rather clear that $\mathcal{S}(a, b) = 0$ if $0 \in (a, b)$. For the calculation of the cohomology of $\mathcal{S}$ on the sector $(-\pi, \pi)$ it suffices to consider the open covering $\{(-\pi, 0), (-\delta, \delta), (0, \pi)\}$, with $\delta > 0$, since the cohomology on $(-\pi, 0)$ and $(0, \pi)$ is trivial. The Čech cohomology of $\mathcal{S}$ with respect to this covering has clearly the property $H^1 = 0$. It follows that $H^1((-\pi, \pi), \mathcal{S}) = 0$. From this we conclude that $\hat{y}$ has an asymptotic lift on the sector $(-\pi, \pi)$. This lift is unique since $\mathcal{S}(-\pi, \pi) = 0$ and therefore equal to v_1. We conclude that the sheaf $\mathcal{S}$ has trivial H^1 for every open subset of $(-\pi, \pi)$. One can express this as: "all the 1-cohomology of the sheaf $\mathcal{S}$ is concentrated in the direction π".

One can also find the asymptotic lift v_2 be a calculation of cohomology. Indeed, by Remarks 11.9 there is an asymptotic lift v of $\hat{y}$ on the sector $(\frac{\pi}{2}, \frac{3\pi}{2})$. Take $\delta > 0$ and small. The interval $(\frac{\pi}{2} - \delta, \frac{3\pi}{2})$ has a covering by $(\frac{\pi}{2}, \frac{3\pi}{2})$ and $(\frac{\pi}{2} - \delta, \pi)$. The H^1 of $\mathcal{S}$ with respect to this covering can be seen to be zero. This proves the existence of v_2.

More generally, let a, b be singular directions, such that (a, b) does not contain any singular direction. Then one can show, using the description above of the sheaf $\mathcal{S}$, that $H^1((a - \frac{\pi}{2}, b + \frac{\pi}{2}), \mathcal{S}) = 0$. This proves the existence of an asymptotic lift on the sector $(a - \frac{\pi}{2}, b + \frac{\pi}{2})$. Since also $H^0((a - \frac{\pi}{2}, b + \frac{\pi}{2}), \mathcal{S}) = 0$, this asymptotic lift is unique and coincides with the multisum with respect to the directions in (a, b).

Chapter 12

q-difference equations

12.1   Formal aspects

We start by analyzing the formal aspect of q-difference equations. We suppose that the number $q \in \mathbf{C}$ is not zero and not a root of unity. A logarithm $2\pi i \tau$ of q is fixed. On the fields

$$\mathbf{C}(z), \ k_\infty := \mathbf{C}(\{z^{-1}\}), \ \hat{k}_\infty := \mathbf{C}((z^{-1})), \ k_0 := \mathbf{C}(\{z\}) \text{ and } \hat{k}_0 := \mathbf{C}((z))$$

there is a natural action of ϕ given by $\phi(z) = qz$. Let K denote the union of the fields $\mathbf{C}(z^{1/m})$ for $m = 1, 2, \ldots$. The symbols z^λ with $\lambda \in \mathbf{Q}$ are chosen such that $z^{\lambda_1} z^{\lambda_2} = z^{\lambda_1 + \lambda_2}$. On the field K one extends the action of ϕ by $\phi(z^\lambda) = e^{2\pi i \tau \lambda} z^\lambda$. On the algebraic closures of $k_0, \hat{k}_0, k_\infty$ and $\hat{k}_\infty$ the extension of the action of ϕ is defined by the same formula. Let k be any of the fields above. A q-difference module over k is a pair (M, Φ) where M is a finite dimensional vector space M over k and $\Phi : M \to M$ is an invertible $\mathbf{C}$-linear map such that $\Phi(fm) = \phi(f)\Phi(m)$ for any $f \in k$ and $m \in M$.

The next step is to find "normal forms" for q-difference modules. Let $\mathcal{P} = \mathcal{P}_\infty$ denote the algebraic closure of $\hat{k}_\infty$. This is the field of formal Puiseux series in the variable z^{-1}. The group of equivalence classes of one dimensional modules is equal to $\mathcal{P}_\infty^*/U$ where U is the subgroup consisting of the elements $\frac{\phi(f)}{f}$ with $f \in \mathcal{P}_\infty^*$. This group U turns out to be

$$\{e^{2\pi i \tau \mu}(1 + d) | \, \mu \in \mathbf{Q}, \, d \in z^{-1/m}\mathbf{C}[[z^{-1/m}]] \text{ for some } m \geq 1.\}$$

The group $\mathcal{P}_\infty^*/U$ is isomorphic with

$$\{z^\lambda | \, \lambda \in \mathbf{Q}\} \times \mathbf{C}^*/\{e^{2\pi i \tau \mu} | \, \mu \in \mathbf{Q}\}.$$

It is not difficult to show that every module over $\mathcal{P}_\infty$ is isomorphic to a direct sum $\sum_{\lambda,c} E(z^\lambda c) \otimes M_{\lambda,c}$ where $E(z^\lambda c)$ denotes the one dimensional module $\mathcal{P}_\infty e_{\lambda,c}$ with $\Phi(e_{\lambda,c}) = z^\lambda c e_{\lambda,c}$ and where each $M_{\lambda,c}$ has a basis such that the matrix of Φ with respect to this basis is a constant unipotent matrix. This representation is not unique. Indeed, $E(e^{2\pi i\tau\mu}) \cong E(1)$ for every rational μ. If one fixes a set of representatives S in $\mathbf{C}^*$ of $\mathbf{C}^*/\{e^{2\pi i\tau\mu} \mid \mu \in \mathbf{Q}\}$ then every q-difference module over $\mathcal{P}_\infty$ is isomorphic to a unique direct sum $\sum_{\lambda \in \mathbf{Q}, c \in S} E(z^\lambda c) \otimes M_{\lambda,c}$.

We want now to construct a *universal Picard-Vessiot ring* over $\hat{k}_\infty$ for the q-difference equations over this field. The difference ring $R := \mathbf{C}(z)[\{e(z^\lambda c)\}_{\lambda \in \mathbf{Q}, c \in \mathbf{C}^*}, l]$ is defined by the following relations:

- $e(z^{\lambda_1}c_1)e(z^{\lambda_2}c_2) = e(z^{\lambda_1+\lambda_2}c_1c_2)$.

- $e(1) = 1$ and $e(q) = z$.

- $\phi(e(z^\lambda c)) = e(e^{2\pi i\tau\lambda})ce(z^\lambda c)$.

- $\phi(l) = l + 1$.

Taking the tensor product with $\hat{k}_\infty$ over $\mathbf{C}(z)$ gives a similar ring

$$\hat{R}_\infty := \hat{k}_\infty[\{e(z^\lambda c)\}, l].$$

We note that $\mathcal{P}_\infty$ embeds in $\hat{R}_\infty$ in a canonical way be sending z^λ to $e(e^{2\pi i\tau\lambda})$. It is not difficult to show that the last ring is the universal Picard-Vessiot ring for q-difference equations over $\hat{k}_\infty$. This means:

1. $\hat{R}_\infty$ has no proper ϕ-invariant ideals.

2. The set of ϕ-invariant elements of $\hat{R}_\infty$ is $\mathbf{C}$.

3. Every q-difference equation over $\hat{k}_\infty$ has a fundamental matrix with coefficients in $\hat{R}_\infty$.

4. No proper difference subring has the properties (1) and (3).

We note that $\hat{R}_\infty$ is also the universal Picard-Vessiot ring for the q-difference equations over $\mathcal{P}_\infty$.

The group of all $\mathbf{C}(z)$-linear automorphisms of $R = \mathbf{C}(z)[\{e(z^\lambda c)\}_{\lambda \in \mathbf{Q}, c \in \mathbf{C}^*}, l]$, which commute with ϕ will be denoted by $\mathbf{G}$. Any element $\sigma \in \mathbf{G}$ can be described by a triple (h, s, a) as follows:
$h : \mathbf{C}^* \to \mathbf{C}^*$ is a homomorphism with $h(q) = 1$, $s : \mathbf{Q} \to \mathbf{C}^*$ is a homomorphism and a is a constant. The action of σ is given by the formulas

$$\sigma(e(z^\lambda c)) = s(\lambda)h(c)e(h(e^{2\pi i\tau\lambda}))e(z^\lambda c) \text{ and } \sigma(l) = l + a.$$

The group $\mathbf{G}$ can be given a natural structure as *affine group scheme* over $\mathbf{C}$. This structure can be defined as follows. Choose an integer $n \geq 1$ and a finitely generated subgroup $C \subset \mathbf{C}^*$, which contains the element $e^{2\pi i \tau / n}$. Let $R_{n,C}$ denote the subring $\mathbf{C}(z)[\{e(z^\lambda c)\}_{\lambda \in 1/n\mathbf{Z}, c \in C}, l]$ of R. The ring $R_{n,C}$ is invariant under ϕ and also invariant under each $\sigma \in \mathbf{G}$. Let $\mathbf{G}_{n,C}$ denote the group of the automorphisms of $R_{n,C}$ which commute with the action of ϕ. Then $\mathbf{G}_{n,C}$ is in an obvious way a linear algebraic group over $\mathbf{C}$. The restriction homomorphism $\mathbf{G} \to \mathbf{G}_{n,C}$ is surjective. Further, R is the filtered union of the $R_{n,C}$. Therefore $\mathbf{G}$ is the projective limit of the groups $\mathbf{G}_{n,C}$. This makes $\mathbf{G}$ into an affine group scheme over $\mathbf{C}$.

The subgroup $\mathbf{G}^o$ of $\mathbf{G}$ consists of the σ's such that the corresponding triple (h, s, a) has the property that the subgroup

$$\{e^{2\pi i (\tau \lambda_1 + \lambda_2)} \mid \lambda_1, \lambda_2 \in \mathbf{Q}\}$$

of $\mathbf{C}^*$ lies in the kernel of h. This group is in a similar way the projective limit of groups $\mathbf{G}_{n,C}^o$. One easily verifies that each $\mathbf{G}_{n,C}^o$ is commutative and connected. In fact $\mathbf{G}_{n,C}^o$ is a product of a torus and the additive group $\mathbf{G}_a$. Further $\mathbf{G}_{n,C}^o$ is the component of the identity of the group $\mathbf{G}_{n,C}$.

Special elements in the group $\mathbf{G}$, which will be used later, are γ and δ defined by:

γ has $s = 1$, $a = 0$ and $h_1 : \mathbf{C}^* \to \mu_\infty \subset \mathbf{C}^*$ a homomorphism such that $h_1(e^{2\pi i \tau \lambda}) = e^{2\pi i \lambda}$ for all $\lambda \in \mathbf{Q}$ and h_1 maps μ_∞ to 1. Here μ_∞ denotes the group of the roots of unity in $\mathbf{C}^*$. The definition of γ is chosen such that $\gamma e(z^\lambda) = e(e^{2\pi i \lambda} z^\lambda)$ and $\gamma e(e^{2\pi i \lambda}) = e(e^{2\pi i \lambda})$ for rational λ.

δ has $s = 1$ and $a = 0$ and $h_2 : \mathbf{C}^* \to \mu_\infty \subset \mathbf{C}^*$ is a homomorphism which is the identity on μ_∞ and maps all $e^{2\pi i \tau \lambda}$ with $\lambda \in \mathbf{Q}$ to 1. The δ is chosen such that $\delta e(z^\lambda) = e(z^\lambda)$ and $\delta e(e^{2\pi i \lambda}) = e^{2\pi i \lambda} e(e^{2\pi i \lambda})$ for rational λ.

The elements γ and δ *do not commute*. The commutator $\epsilon := \delta \gamma \delta^{-1} \gamma^{-1}$ has the properties:

$$\epsilon e(c) = e(c), \quad \epsilon e(z^\lambda) = e^{2\pi i \lambda} e(z^\lambda), \quad \epsilon \in \mathbf{G}^o, \quad \epsilon \text{ commutes with } \delta \text{ and } \gamma.$$

An analysis of the actions of γ and δ shows that the only relations in the group generated by γ and δ are $\epsilon \gamma = \gamma \epsilon$ and $\epsilon \delta = \delta \epsilon$. This implies that the group generated by γ and δ is isomorphic to the Heisenberg group, i.e. the group

$$\left\{ \begin{pmatrix} 1 & a & b \\ 0 & 1 & c \\ 0 & 0 & 1 \end{pmatrix} \mid a, b, c \in \mathbf{Z} \right\}.$$

The group $\mathbf{G}^o$ is a normal subgroup of $\mathbf{G}$. The group $\mathbf{G}/\mathbf{G}^o$ can be identified with the homomorphisms of the torsion subgroup $Tors$ of $\mathbf{C}^*/\{q^n \mid n \in \mathbf{Z}\}$ to $\mathbf{C}^*$. The group $Tors$ is isomorphic to the product of two copies of $\mathbf{Q}/\mathbf{Z}$. The images of γ and δ in $\mathbf{G}/\mathbf{G}^o$ are "topological generators" of this group. This has the following meaning: For every finitely generated subgroup K of $Tors$, the group of homomorphisms of K to $\mathbf{C}^*$ is generated by the images of γ and δ. This statement follows at once from the definitions of γ and δ.

The q-difference module M over $\hat{k}_\infty$ is called *regular singular* if the fundamental matrix for M does not involve the terms $e(z^\lambda)$ with $\lambda \neq 0$. There are several ways to reformulate that property. The following statements on M are equivalent:

1. M is regular singular.

2. There is a complex vector space $V \subset M$ such that V is invariant under the action of Φ and such that the canonical map $\hat{k}_\infty \otimes V \to M$ is an isomorphism.

3. The module M has a matrix representation $y(qz) = A\, y(z)$ with a constant invertible matrix A.

The module M over $\hat{k}_\infty$ is called *regular* if M is trivial, i.e. $M \cong \hat{k}_\infty^n$ with the ordinary action of ϕ.

The group of the automorphisms of $\hat{R}_\infty$ which are the identity on $\hat{k}_\infty$ and commute with ϕ is identical with the group $\mathbf{G}$ that we have just described. For a difference module M over $\hat{k}_\infty$ one defines $\omega_\infty(M) = ker(\Phi - 1, \hat{R}_\infty \otimes M)$. This is the vector space of "the solutions of the equation". The group $\mathbf{G}$ acts on $\omega_\infty(M)$. The image of this group is the difference Galois group G of M. The structure of the group scheme $\mathbf{G}$ implies the following results:

Proposition 12.1 *Let G be the difference Galois group of a q-difference module M over $\hat{k}_\infty$. Then G^o, the component of the identity of G, is either a torus or the product of a torus and the additive group $\mathbf{G}_a$. The group G/G^o is commutative and is generated by at most two elements.*
Suppose that M is regular singular. Then G is commutative. The subgroup $H \subset G$, generated by the images of γ and δ, is a finite commutative subgroup which maps surjectively to G/G^o.

For q-difference modules over $\hat{k}_0$ the situation is analogous.

Proposition 12.2 *Let G denote the difference Galois group G of a q-difference module M over $\mathbf{C}(z)$.*

1. *The group G/G^o is commutative and is generated by at most two elements.*

2. *Suppose that $\mathring{k}_\infty \otimes M$ or $\mathring{k}_0 \otimes M$ is regular singular. There is a finite commutative subgroup H, with at most two generators, such that the map $H \to G/G^0$ is surjective.*

3. *Suppose that $\mathring{k}_\infty \otimes M$ or $\mathring{k}_0 \otimes M$ is regular then the difference Galois group of M is connected.*

Proof: (1) The embedding $\mathbf{C}(z) \subset \mathring{k}_\infty$ induces an inclusion of the difference Galois group G_∞ of $\mathring{k}_\infty \otimes M$ into G. Let G' denote the subgroup of G generated by G^o and the images of γ and δ. Since $G^o_\infty \subset G^o$ and because G_∞ is generated by G^o_∞ and the images of γ and δ, one has $G_\infty \subset G'$. The Picard-Vessiot ring of M is denoted by PV. One can see PV as a subring of the Picard-Vessiot ring PV' of the module $\mathring{k}_\infty \otimes M$. For the invariants one has the following inclusion $PV^{G'} \subset (PV')^{G_\infty} = \mathring{k}_\infty$. Since $PV^{G'}$ is a finite extension of $\mathbf{C}(z)$ one concludes that $PV^{G'}$ is a field. The existence of a ϕ-action on $PV^{G'}$ implies that $PV^{G'} = \mathbf{C}(z^{1/m})$ for some $m \geq 1$. From $PV^{G'} \subset \mathring{k}_\infty$ it follows that $PV^{G'} = \mathbf{C}(z)$. This implies that $G' = G$. Hence G/G^o is generated by the images of γ and δ. The commutator of the images of γ and δ belongs to $G^o_\infty \subset G^o$. This proves the statement.

(2) If $\mathring{k}_\infty \otimes M$ is regular singular then the images of γ and δ in $G_\infty \subset G$ generate a finite commutative subgroup H of $G_\infty \subset G$. The map $H \to G/G^o$ is surjective.

(3) If $\mathring{k}_\infty \otimes M$ is regular then $G_\infty = 1$ and the group H of (2) is the trivial group. This implies that G is connected. ∎

12.2 Analytic properties

We start the analytic part by considering q-difference equations over k_0. The case k_∞ is of course quite the same. The following example shows that we have to assume that $|q| < 1$ (or $|q| > 1$) in order to obtain a reasonable analytic theory.

Example 12.3 *The equation $y(qz) = (1 + z)y(z)$ over k_0*

The coefficients of the unique formal solution $y = \sum_{n \geq 0} a_n z^n$ with $a_0 = 1$ are given by the recurrence relation $(q^n - 1)a_n = a_{n-1}$.
Hence $a_n = (q^n - 1)^{-1} \ldots (q - 1)^{-1}$. Suppose that $|q| = 1$ and so $\tau \in \mathbf{R} \setminus \mathbf{Q}$. One has $|1 - q^n|^2 = 2(1 - \cos(2\pi n \tau))$. If the formal series y is convergent then there is a positive r with $1 - \cos(2\pi n \tau) \geq r^{2n}/2$ for all $n \geq 1$. This implies $\inf |n\tau + \mathbf{Z}| \geq r^n$ for all $n \geq 1$. In other words $|\tau + \frac{m}{n}| \geq \frac{r^n}{n}$ for every integer m (and also for every $n \geq 1$). Liouville numbers can be rapidly approximated by rational numbers. If τ is a Liouville number then the divergence of the series is

"chaotic" in the sense that there is no hope that the formal series is summable. For real algebraic, non rational τ one has the famous Roth's inequality

$$|\tau + \frac{m}{n}| \geq \frac{c(\tau, \epsilon)}{n^{2+\epsilon}}$$

with a positive constant $c(\tau, \epsilon)$ depending on τ and ϵ. This inequality assures that the series y is convergent. These are (as far as we know) the only concrete examples of real numbers which cannot be too rapidly approximated by rational numbers. One knows however that this subset of the real numbers has full measure.

If $|q| \neq 1$ then the series y is evidently convergent.

In the sequel we will assume that $|q| < 1$ or $Im(\tau) > 0$.

12.2.1   Regular singular equations over k_0

By *regular singular equation M over k_0* we mean that $\hat{k}_0 \otimes_{k_0} M$ is regular singular. One easily sees that this is again equivalent to:

M has a matrix representation $y(qz) = A\, y(z)$ with $A \in Gl(n, \mathbf{C}\{z\})$.

Let $y(qz) = A\, y(z)$ be difference equation in matrix form with $A = A_0 + A_1 z + \ldots \in Gl(n, \mathbf{C}\{z\})$. Let $\alpha_1, \ldots, \alpha_r$ denote the eigenvalues of A_0. One may further suppose that the matrix A_0 is in upper triangular form and that the first diagonal entry of A_0 is α_1. Let F denote the diagonal matrix with diagonal entries $z^{-1}, 1, \ldots, 1$. The transformed matrix $\tilde{A} := F(qz)^{-1}AF(z)$ is again in $Gl(n, \mathbf{C}\{z\})$ and the eigenvalues of $\tilde{A}_0$ are $q\alpha_1, \alpha_2, \ldots, \alpha_r$. A similar process, which changes the eigenvalue α_r into $q^{-1}\alpha_r$, can be applied to A. Repeating this process one finally finds an equation $y(qz) = B\, y(z)$ which is equivalent to $y(qz) = A\, y(z)$ such that the eigenvalues β_j of B_0 satisfy $|q| < |\beta_j| \leq 1$.

The next result shows that regular singular equations are rather simple and do not give rise to problems about asymptotic expansions.

Lemma 12.4 *Let $A = A_0 + A_1 z + \ldots$ be an invertible matrix with coefficients in $\mathbf{C}\{z\}$. Suppose that eigenvalues $\alpha_1, \ldots, \alpha_r$ of A_0 have the property that $|q| < |\alpha_j| \leq 1$. The unique formal matrix $F = 1 + F_1 z + F_2 z^2 + \ldots$ with $F(qz)^{-1}AF(z) = A_0$ is convergent.*

Proof: For the constant matrices F_n there is the following recurrence relation

$$q^n F_n A_0 - A_0 F_n = A_1 F_{n-1} + \ldots + A_{n-1} + A_n.$$

The map $L_n : B \mapsto q^n B A_0 - A_0 B$, $n \geq 1$, from the vector space of the matrices to itself is bijective. Indeed, if $L_n(B) = 0$ then $A_0 B A_0^{-1} = q^n B$. The eigenvalues of the map $B \mapsto A_0 B A_0^{-1}$ are the $\alpha_i \alpha_j^{-1}$. Hence $B = 0$. This shows the existence and uniqueness of the formal matrix $\hat{F}$. Since $|q|^n$ tends to 0, there is a constant c such that the norms of the inverse maps L_n^{-1} is bounded by c. Using this one easily proves that F is convergent.   ∎

As a consequence one obtains that every regular singular difference equation over k_0 is equivalent with an equation $y(qz) = A\, y(z)$ where A is a constant invertible matrix. One can moreover specify that the eigenvalues c of A satisfy $|q| < |c| \leq 1$. With this specification, the constant matrix A is unique up to conjugation in $Gl(n, \mathbf{C})$.

A translation of the above for regular singular modules M is the following: Let

$$\omega_0(M) = ker(\Phi - 1, k_0[\{e(c)\}_{c \in \mathbf{C}^*}, l] \otimes M).$$

Then $\omega_0(M)$ is a finite dimensional vector space over $\mathbf{C}$ and the canonical map

$$k_0[\{e(c)\}_{c \in \mathbf{C}^*}, l] \otimes \omega_0(M) \to k_0[\{e(c)\}_{c \in \mathbf{C}^*}, l] \otimes_{k_0} M$$

is an isomorphism. An automorphism σ of $k_0[\{e(c)\}_{c \in \mathbf{C}^*}, l]$ which is the identity on k_0 and commutes with ϕ is determined by a homomorphism $h : \mathbf{C}^* \to \mathbf{C}^*$ such that $h(q) = 1$ and an element $a \in \mathbf{C}$. The action of σ is $\sigma(e(c)) = h(c)e(c)$ for all $c \in \mathbf{C}^*$ and $\sigma(l) = l + a$. This group is commutative and has the structure of an affine group scheme over $\mathbf{C}$. As before, the group is topologically generated by two special elements γ and δ and a connected subgroup corresponding to the the homomorphisms h which are trivial on

$$\{e^{2\pi i(\tau \lambda_1 + \lambda_2)} | \lambda_1, \lambda_2 \in \mathbf{Q}\}.$$

Consequently, the difference Galois group G of any regular singular equation over k_0 has the properties:

- G^o is a torus or the product of a torus and the additive group $\mathbf{G}_a$.

- There is a finite commutative subgroup $H \subset G$ with at most two generators such that $H \to G/G^o$ is surjective.

- G is commutative.

This finishes the theory of regular singular q-difference equations over the fields k_0 and k_∞.

Remarks 12.5 *More general equations over k_0*

(1) We give an example to show that formal solutions are in general not multisummable. Consider the equation

$$y(qz) = \begin{pmatrix} z & 1 \\ 0 & 1 \end{pmatrix} y(z).$$

There is a unique formal matrix $\begin{pmatrix} 1 & f \\ 0 & 1 \end{pmatrix}$ with $F(qz)^{-1}AF(z) = \begin{pmatrix} z & 0 \\ 0 & 1 \end{pmatrix}$.

The element f turns out to be $f = \sum_{n>0} q^{-\frac{n^2+n}{2}} z^n$. This series is divergent and not multisummable since for a multisummable series $\sum a_n z^n$ there is an inequality of the form $|a_n| \le c_1 (n!)^{c_2}$ for positive constants c_1, c_2. The classical methods for the interpretation of the formal series f do not work. It is possible that a quite different summation method can be applied in order to give the formal f some meaning. See also [52]

(2) Let us suppose for an instant that $|q| = 1$ and that $\tau \in \mathbf{R}$ cannot be too well approximated by rational numbers. In that case one can show that *every* difference equation over k_0 is equivalent (over the algebraic closure of k_0) with an equation $y(qz) = A\, y(z)$, where A is a direct sum of blocks of the form $z^{\lambda_i} A_i$ with rational λ_i and constant invertible matrices A_i. This means that the analytic theory of equations over k_0 coincides with the formal theory over $\hat{k}_0$. This looks quite interesting. The disadvantage of the choice of a q with $|q| = 1$ becomes apparent if one studies q-difference equation $y(qz) = A\, y(z)$ over $\mathbf{C}(z)$. Suppose for simplicity that $A(0) = A(\infty) = 1$. The equation has a fundamental matrix $F_0 \in Gl(n, \mathbf{C}\{z\})$ at 0 and also a holomorphic fundamental matrix F_∞ at ∞. The fundamental matrix F_0 is defined in some neighborhood of 0. The equation $F_0(qz) = A(z)F_0(z)$ cannot be used to extend F_0 to the entire complex plane since $|q| = 1$. One is not able to compare the two fundamental matrices and no information about the difference Galois group of the equation over $\mathbf{C}(z)$ is obtained. This is the second reason why we will take a q with $|q| < 1$ in the sequel.

12.2.2   Equations over $\mathbf{C}(z)$

We continue with the investigation of certain difference modules M over $\mathbf{C}(z)$, namely the regular singular ones. A difference module M is called *regular singular* if both $k_0 \otimes M$ and $k_\infty \otimes M$ are regular singular. The module M is called *regular* if $k_0 \otimes M$ and $k_\infty \otimes M$ are trivial difference modules. We start with an example which indicates the general method that we will develop.

Example 12.6 $y(qz) = \frac{(z-a)(z-a^{-1})}{(z-1)^2} y(z)$ *with* $a \in \mathbf{C}^*$

This is a regular order one equation over $\mathbf{C}(z)$. It has local solutions y_0 and y_∞ at 0 and ∞. The local solutions are normalized by $y_0(0) = 1$ and

$y_\infty(\infty) = 1$. The solution y_0 is a meromorphic function on the entire complex plane. The solution y_∞ is meromorphic on $\mathbf{C}^* \cup \{\infty\}$. The function $f := y_0^{-1} y_\infty$ is meromorphic on $\mathbf{C}^*$ and q-invariant. Hence f is a meromorphic function on the elliptic curve $E =: \mathbf{C}^*/q^{\mathbf{Z}}$. One can calculate the poles and the zeros of f as follows. The function y_∞ has neither poles nor zeroes in a neighborhood of ∞. The equation $y_\infty(qz) = \frac{(z-a)(z-a^{-1})}{(z-1)^2} y_\infty(z)$ shows that the divisor of y_∞ is equal to

$$\sum_{n \geq n_1} [q^n a] + \sum_{n \geq n_2} [q^n a^{-1}] + \sum_{n \geq n_3} -2[q^n] \text{ for certain } n_1, n_2, n_3.$$

The function y_0^{-1} has no zeros or poles in a neighborhood of 0. It follows that the divisor of f on E is $[\rho(a)] + [\rho(a^{-1})] - 2[\rho(1)]$, where $\rho : \mathbf{C}^* \to E$ is the canonical map. One conclusion is that f is constant if and only if a is an integral power of q. Further f is constant if and only if the equation has a non-trivial solution in $\mathbf{C}(z)$. The choice $a = -1$ leads to a function f with divisor $2[\rho(-1)] - 2[\rho(1)]$. Hence for this choice, f is the normalized Weierstrass function on E.
One can give explicit expressions for y_0 and y_∞, namely

$$y_0 = \prod_{j \geq 0} \left(\frac{(q^j z - a)(q^j z - a^{-1})}{(q^j z - 1)^2} \right)^{-1} \text{ and } y_\infty = \prod_{j \geq 1} \frac{(q^{-j} z - a)(q^{-j} z - a^{-1})}{(q^{-j} z - 1)^2}$$

Using this product one finds a product formula for f. This formula is of course the usual way to write a meromorphic function on E as a product of theta functions.

Remarks: In the recent paper [22] the Galois group of a matrix equation $y(qz) = A(z)y(z)$ with $A(0) = A(\infty) = 1$ is studied. This is a regular equation. For a matrix equation as above, P.I. Etingof defines a connection matrix and shows that this matrix "generates" the difference Galois group of the equation. Our methods are somewhat different and extend to the larger class of regular singular equations.

12.3 Construction of the connection map

12.3.1 Meromorphic vector bundles

We will need some less known properties of meromorphic matrices. We start with some definitions. For any connected Riemann surface X, we write $\mathcal{O}_X = \mathcal{O}$ and $\mathcal{M}_X = \mathcal{M}$ for the sheaves of holomorphic and meromorphic functions on X. A *holomorphic vector bundle of rank n* is a sheaf of $\mathcal{O}$-modules which is locally isomorphic to $\mathcal{O}^n$. A *meromorphic vector bundle of rank n* on X is a sheaf of $\mathcal{M}$-modules which is locally isomorphic to $\mathcal{M}^n$. A holomorphic vector bundle can be described by an element in $H^1(X, Gl(n, \mathcal{O}))$ and a meromorphic bundle by an element of $H^1(X, Gl(n, \mathcal{M}))$. The following lemma says that every meromorphic vector bundle comes from a holomorphic vector bundle. We give here essentially the proof of C. Praagman [48].

Lemma 12.7 *The canonical map $H^1(X, Gl(n, \mathcal{O})) \to H^1(X, Gl(n, \mathcal{M}))$ is surjective.*

Proof: Let L be a meromorphic vector bundle represented by an element $\xi \in H^1(X, Gl(n, \mathcal{M}))$. This element can be calculated with Čech cohomology. The space X is paracompact and there are two locally finite open coverings $U := \{U_i\}$ and $V := \{V_i\}$ of X such that:

1. ξ is represented by elements $A_{i,j} \in Gl(n, \mathcal{M}(U_i \cap U_j))$.

2. V_i is a relatively compact subset of U_i.

The set of points $S_{i,j} \subset V_i \cap V_j$, where $A_{i,j}$ is not an invertible holomorphic matrix, is finite. Since the covering V is locally finite the set $S := \cup_{i,j} S_{i,j}$ is a discrete subset of X. Let $V_i^* = V_i \setminus S$ and $X^* = X \setminus S$. Then $\{V_i^*\}$ is an open covering of X^* and the restriction of the 1-cocycle $\{A_{i,j}\}$ to $V_i^* \cap V_j^*$ has values in $Gl(n, \mathcal{O})$. The corresponding holomorphic vector bundle L_0 has the property that the natural morphism $f : \mathcal{M} \otimes_{\mathcal{O}} L_0 \to L|_{X^*}$ is an isomorphism. (Here $L|_{X^*}$ denotes the restriction of L to X^*.) The holomorphic vector bundle L_0 can be extended to X since S is a discrete set. This extension, again called L_0, can be made such that L_0 is a subsheaf of L. The morphism f extends then in a unique way to an isomorphism $f : \mathcal{M} \otimes_{\mathcal{O}} L_0 \to L$. ∎

Corollary 12.8

1. Any meromorphic vector bundle on X is trivial.

2. For every $T \in Gl(n, \mathcal{M}(\mathbf{C}^*))$ there are $F_0 \in Gl(n, \mathcal{M}(\mathbf{C}))$ and $F_\infty \in Gl(n, \mathcal{M}(\mathbf{C}^* \cup \{\infty\}))$ with $F_0^{-1} F_\infty = T$.

Proof: 1. The meromorphic vector bundle L comes from some holomorphic bundle L_0 on X. If X is an open Riemann surface then every holomorphic vector bundle on X is trivial according to Theorem 30.4 of [24]. If X is compact then L_0 "is" an algebraic bundle on the connected projective non singular algebraic curve X. Such a bundle trivializes over the function field of X (Corollary 29.17 of [24]).

2. Using the covering of the projective line $\mathbf{P}_{\mathbf{C}}^1$ over $\mathbf{C}$ by the two sets $\mathbf{C}$ and $\mathbf{C}^* \cup \{\infty\}$, one can view the matrix T as a 1-cocycle for $Gl(n, \mathcal{M})$. This 1-cocycle is trivial since $H^1(X, Gl(n, \mathcal{M}))$ is trivial. This proves the existence of F_0 and F_∞. ∎

12.3.2 The connection map of a regular equation

In the sequel we will use the following notation: $\mathcal{M}(X)$ is the field of meromorphic functions on any connected Riemann surface X. Let E denote the elliptic curve $\mathbf{C}^*/q^{\mathbf{Z}}$ (which is isomorphic to $\mathbf{C}/\mathbf{Z}\tau + \mathbf{Z}$). We note that the field of meromorphic functions $\mathcal{M}(E)$ coincides with the set of the ϕ-invariant elements of $\mathcal{M}(\mathbf{C}^*)$

Our aim is to construct a connection map for regular singular modules over $\mathbf{C}(z)$ and to give its relation with the difference Galois group. We start with regular modules M in order to illustrate the main ideas.

Put $\omega_0(M) = ker(\Phi - 1, \mathcal{M}(\mathbf{C}) \otimes M)$ and

$$\omega_\infty(M) = ker(\Phi - 1, \mathcal{M}(\mathbf{C}^* \cup \{\infty\}) \otimes M).$$

From the local theory at 0 it follows that the canonical map

$$\mathcal{M}(\mathbf{C}) \otimes_{\mathbf{C}} \omega_0(M) \to \mathcal{M}(\mathbf{C}) \otimes_{\mathbf{C}(z)} M$$

is an isomorpism. A similar result holds for $\omega_\infty(M)$. The two spaces $\omega_0(M)$ and $\omega_\infty(M)$ have natural maps to $\omega_*(M) := ker(\Phi - 1, \mathcal{M}(\mathbf{C}^*) \otimes M)$ and one can compare their images. Roughly speaking, this will produce the "connection matrix". The subset of ϕ-invariants elements in $\mathcal{M}(\mathbf{C}^*)$ is equal to $\mathcal{M}(E)$. The map $\omega_0(M) \to \omega_*(M)$ induces an isomorphism $\mathcal{M}(E) \otimes_{\mathbf{C}} \omega_0(M) \to \omega_*(M)$. There is a similar isomorphism $\mathcal{M}(E) \otimes_{\mathbf{C}} \omega_\infty(M) \to \omega_*(M)$. From this one finds an isomorphism

$$S_M : \mathcal{M}(E) \otimes \omega_0(M) \to \mathcal{M}(E) \otimes_{\mathbf{C}} \omega_\infty(M).$$

The map S_M will be called the *connection map* (or matrix) of M. For any regular module M we associate the triple $(\omega_0(M), \omega_\infty(M), S_M)$. We note that, unlike the connection maps defined for ordinary difference equations, this connection map relates different vector spaces.

Let us introduce a category **Triples** of triples. An object of this category is a triple (V_0, V_∞, S), where V_0, V_∞ are two vector spaces over $\mathbf{C}$ with the same finite dimension and where $S : \mathcal{M}(E) \otimes V_0 \to \mathcal{M}(E) \otimes V_\infty$ is a $\mathcal{M}(E)$-linear isomorphism. A morphism $f = (f_0, f_\infty)$ from the triple (V_0, V_∞, S) to the triple (V_0', V_∞', S') is a pair of **C**-linear maps $f_0 : V_0 \to V_0'$ and $f_\infty : V_\infty \to V_\infty'$ such that $S'f_0 = f_\infty S$. The tensor product of the two triples (V_0, V_∞, S) and (V_0', V_∞', S') is defined as $(V_0 \otimes V_0', V_\infty \otimes V_\infty', S \otimes S')$. With those definitions **Triples** is in an obvious way a **C**-linear tensor category. The forgetful functor $\eta : \textbf{Triples} \to Vect_{\mathbf{C}}$ to the tensor category $Vect_{\mathbf{C}}$ of the finite dimensional vector spaces over **C** is a fibre functor. Hence **Triples** is a neutral **C**-linear

Tannakian category. The main result on regular q-difference equations over $\mathbf{C}(z)$ is the following theorem.

Theorem 12.9

(a) The map $M \mapsto (\omega_0(M), \omega_\infty(M), S_M)$ from the category of regular difference modules over $\mathbf{C}(z)$ to the category **Triples** is an equivalence of $\mathbf{C}$-linear tensor categories.

(b) The difference Galois group G of the regular q-difference module M over $\mathbf{C}(z)$, seen as a subgroup of $Gl(\omega_0(M))$, is the smallest algebraic subgroup which contains $S_M^{-1}(a)S_M(b)$ for all $a, b \in E$ such that $S_M(a)$ and $S_M(b)$ are invertible matrices.

(c) For every connected linear group G there is a regular q-difference module with difference Galois group G.

Proof: (a) The functor $\mathcal{F}$, indicated in the statement of (a), is $\mathbf{C}$-linear and preserves all constructions of linear algebra (in particular tensor products). $\mathcal{F}$ is fully faithful if the map $\mathrm{Hom}(1, M) \to \mathrm{Hom}(\mathcal{F}1, (\omega_0(M), \omega_\infty(M), S_M))$ is bijective. The lefthand side consists of the elements $m \in M$ with $\phi(m) = m$. The object $\mathcal{F}1$ is equal to $(\mathbf{C}e_0, \mathbf{C}e_\infty, T)$ with $Te_0 = e_\infty$. Hence the righthand side is equal to the set of pairs $(v_0, v_\infty) \in \omega_0(M) \times \omega_\infty(M)$ with $S_M v_0 = v_\infty$. From this description it is clear that the map is a bijection. We are left with proving that any object (V_0, V_∞, T) of **Triples** is isomorphic to some $\mathcal{F}M$.

Put $M_0 = \mathcal{M}(\mathbf{C}) \otimes V_0$ and consider the ϕ-action on this module defined by $\phi f(z) \otimes v = f(qz) \otimes v$. Similarly one defines the difference module $M_\infty = \mathcal{M}(\mathbf{C}^* \cup \{\infty\}) \otimes V_\infty$. The two q-difference modules are defined above $\mathbf{C}$ and $\mathbf{C}^* \cup \{\infty\}$. The two modules are glued over the open subset $\mathbf{C}^*$ by the isomorphism $\mathcal{M}(\mathbf{C}^*) \otimes M_0 \to \mathcal{M}(\mathbf{C}^*) \otimes M_\infty$ obtained from T by extending the "scalars" from $\mathcal{M}(E)$ to $\mathcal{M}(\mathbf{C}^*)$. The result is a meromorphic vector bundle on $\mathbf{P}^1(\mathbf{C})$. The isomorphism, induced by T, is equivariant with respect to the two actions of ϕ. Hence we found a meromorphic vector bundle over $\mathbf{P}^1(\mathbf{C})$ together with a ϕ-action. According to 12.3.1 this vector bundle is trivial. Let M denote the set of meromorphic sections of the bundle. Then M is a vector space over $\mathbf{C}(z)$. Any $m \in M$ induces restrictions $m_0 \in M_0$ and $m_\infty \in M_\infty$. The two elements $\phi(m_0)$ and $\phi(m_\infty)$ coincide above $\mathbf{C}^*$. Therefore $\phi(m_0)$ and $\phi(m_\infty)$ glue to an element in M which will be denoted by $\phi(m)$. This defines ϕ on M. The verification that $(\omega_0(M), \omega_\infty(M), S_M)$ is isomorphic to (V_0, V_∞, T) is straightforward.

(b) Consider an object (V_0, V_∞, T) of **Triples**. It generates a full tensor subcategory $\{\{(V_0, V_\infty, T)\}\}$ of **Triples**. The functor $\eta : \{\{(V_0, V_\infty, T)\}\} \to Vect_\mathbf{C}$, given by $\eta(W_0, W_\infty, S) = W_0$, is a fibre functor. The category is therefore a neutral Tannakian category and is isomorphic with the category of the finite dimensional representations of some linear algebraic group G, in fact a subgroup of

$\mathrm{Aut}(V_0)$, which represents the functor $\mathrm{Aut}^\otimes(\eta)$. For $a, b \in E$ such that $T(a)$ and $T(b)$ are invertible maps we will produce an element $\lambda \in \mathrm{Aut}^\otimes(\eta)(\mathbf{C}) = G(\mathbf{C})$. By definition (see [20]) λ is a family $\{\lambda_{(W_0, W_\infty, S)}\}$, where (W_0, W_∞, S) runs over the set of objects of $\{\{(V_0, V_\infty, T)\}\}$. One defines $\lambda_{(W_0, W_\infty, S)} = S^{-1}(a)S(b)$. Since (W_0, W_∞, S) is obtained from (V_0, V_∞, T) by some linear constructions (i.e. tensor products, duals, subquotients) the maps $S(a)$ and $S(b)$ are invertible and the family $\{\lambda_{(W_0, W_\infty, S)}\}$ thus defined has the required properties. Let H denote the smallest algebraic subgroup of $\mathrm{Aut}(V_0)$ which contains all $T^{-1}(a)T(b)$, with $a, b \in E$, for which $T(a)$ and $T(b)$ are invertible maps. Then one finds a functor of tensor categories $\{\{(V_0, V_\infty, T)\}\} \to Repr_H$, where $Repr_H$ is the neutral Tannakian category of the finite dimensional representations of H. It is an exercise to show that this functor is an isomorphism of tensor categories. This shows that $G = H$. Using the isomorphism of (a) one finds the result (b).

(c) Let a connected linear algebraic group G be given as a subgroup of some $\mathrm{Aut}(V_0)$. Choose $V_\infty := V_0$. The field $\mathcal{M}(E)$ contains a subfield $\mathbf{C}(t)$ with transcendental t. According to proposition 3.3 there is some $T \in \mathrm{Aut}(\mathbf{C}(t) \otimes V_0)$ with $T(0) = 1$ and G is the smallest group with $T \in G(\mathbf{C}(t))$. One can see T also an element of $\mathrm{Aut}(\mathcal{M}(E) \otimes V_0)$ and it has the required property. $\blacksquare$

Remarks 12.10 *Regular equations in matrix form.*

If the regular module M happens to have a basis such that the corresponding matrix equation $y(qz) = A\, y(z)$ satisfies $A(0) = A(\infty) = 1$ then one has another way of introducing the "connection matrix". (See [22]).
The fundamental matrix F_0 at 0 is chosen such that $F_0(0) = 1$ and similarly, the fundamental matrix F_∞ is chosen with $F_\infty(\infty) = 1$. One easily proves the formulas

$$F_0 = \prod_{n \geq 0} A(q^n z)^{-1} = A(z)^{-1} A(qz)^{-1} \ldots \text{and}$$

$$F_\infty = \prod_{n > 0} A(q^{-n} z) = A(q^{-1} z) A(q^{-2} z) \ldots.$$

The matrix $T = F_0^{-1} F_\infty$ is equal to $\prod_{n=-\infty}^{\infty} A(q^{-n} z)$. Let s_M denote a matrix for S_M with respect to some basis of $\omega_0(M)$ and $\omega_\infty(M)$. It is not difficult to see that $s_M = CTD$ for some constant invertible matrices C, D. The algebraic subgroup of $Gl(n, \mathbf{C})$ generated by the $s_M^{-1}(a)s_M(b)$ is conjugated to the algebraic subgroup generated by the $T^{-1}(a)T(b)$. The last group is the group appearing in the paper [22]. $\blacksquare$

12.3.3   The connection map of a regular singular equation

We will now consider a *general regular singular difference* module M over $\mathbf{C}(z)$. Let $\omega_0(M)$ denote

$$ker(\Phi - 1, \mathcal{M}(\mathbf{C})[\{e(c)\}_{c \in \mathbf{C}^*}, l] \otimes M).$$

From the local theory at 0 we know that the canonical map

$$\mathcal{M}(\mathbf{C})[\{e(c)\}_{c \in \mathbf{C}^*}, l] \otimes_{\mathbf{C}} \omega_0(M) \to \mathcal{M}(\mathbf{C})[\{e(c)\}_{c \in \mathbf{C}^*}, l] \otimes_{\mathbf{C}(z)} M$$

is an isomorphism. Let $\mathbf{G_{rs}}$ denote the group of the authomorphisms of $\mathbf{C}(z)[\{e(c)\}_{c \in \mathbf{C}^*}, l]$, which are $\mathbf{C}(z)$-linear and commute with ϕ. As in 12.1, this is an affine group scheme. It is commutative. In fact $\mathbf{G_{rs}}$ is a quotient of the affine group scheme $\mathbf{G}$ studied in 12.1. The group $\mathbf{G_{rs}}$ acts on $\omega_0(M)$. One defines $\omega_\infty(M)$ in a similar way. It has analogous properties.

We consider also the difference ring $\mathcal{M}_* := \mathcal{M}(\mathbf{C}^*)[\{e(c)\}, l]$. The group $\mathbf{G_{rs}}$ acts as group of automorphisms of this difference ring. Its subring of ϕ-invariant elements will be denoted by $\mathcal{M}_*^\phi$. Since the action of $\mathbf{G_{rs}}$ commutes with ϕ the group $\mathbf{G_{rs}}$ also acts on $\mathcal{M}_*^\phi$. Let $\omega_*(M)$ denote

$$ker(\Phi - 1, \mathcal{M}(\mathbf{C}^*)[\{e(c)\}, l] \otimes M).$$

This is a module over the ring $\mathcal{M}_*^\phi$ and has a $\mathbf{G_{rs}}$-action. For $f \in \mathcal{M}_*^\phi$, $m \in \omega_*(M)$ and $\sigma \in \mathbf{G_{rs}}$ one has $\sigma(fm) = \sigma(f)\sigma(m)$. There is an obvious $\mathbf{C}$-linear and $\mathbf{G_{rs}}$-equivariant map from $\omega_0(M)$ to $\omega_*(M)$. This map extends as a $\mathbf{G_{rs}}$-equivariant map $\mathcal{M}_*^\phi \otimes_{\mathbf{C}} \omega_0(M) \to \omega_*(M)$, where the action of $\sigma \in \mathbf{G_{rs}}$ on the first term is defined by the formula $\sigma(f \otimes m) = \sigma(f) \otimes \sigma(m)$ for $f \in \mathcal{M}_*^\phi$ and $m \in \omega_0(M)$. We want to show:

Lemma 12.11 *The induced $\mathcal{M}_*^\phi$-linear and $\mathbf{G_{rs}}$-equivariant map*

$$\mathcal{M}_*^\phi \otimes_{\mathbf{C}} \omega_0(M) \to \omega_*(M)$$

is an isomomorphism.

Proof: We choose a matrix equation $y(qz) = A\ y(z)$ with $A \in Gl(n, \mathbf{C}(z))$ representing the module M. Let $F_0 \in Gl(n, \mathcal{M}(\mathbf{C}))$ satisfy $F_0(qz)^{-1}AF_0(z) = C$ where C is a constant matrix in Jordan normal form and eigenvalues $c_1, \ldots, c_n$. Write $C = C_{ss}exp(N)$ with C_{ss} a diagonal matrix with diagonal entries $c_1, \ldots, c_n$ and N a nilpotent matrix commuting with C_{ss}. A basis for $\omega_0(M)$ are the columns of the matrix $G_0 := F_0e(C_{ss})exp(lN)$. Here $e(C_{ss})$ denotes the diagonal matrix with the entries $e(c_1), \ldots, e(c_n)$ on the diagonal. These columns remain unchanged when one considers their image in $\mathcal{M}_* \otimes M$. The matrix is invertible over $\mathcal{M}_*$ and so the map

$$\mathcal{M}_* \otimes \omega_0(M) \to \mathcal{M}_* \otimes M \text{ is a bijection.}$$

Any solution y with coordinates in $\mathcal{M}_*$ of the equation $y(qz) = A\, y(z)$ has the form $G_0 v$ where v is a vector with coordinates in $\mathcal{M}_*^\phi$. This shows that

$$\mathcal{M}_*^\phi \otimes_{\mathbf{C}} \omega_0(M) \to \omega_*(M) \text{ is an isomorphism.}$$

∎

There is also an isomorphism $\mathcal{M}_*^\phi \otimes_{\mathbf{C}} \omega_\infty(M) \to \omega_*(M)$. The two resulting isomorphism

$$f_0, f_\infty : \mathcal{M}_*^\phi \otimes_{\mathbf{C}} \omega(M) \to \omega_*(M)$$

induce a $\mathcal{M}_*^\phi$-linear and $\mathbf{G}_{\mathbf{rs}}$-equivariant isomorphism

$$S_M : \mathcal{M}_*^\phi \otimes_{\mathbf{C}} \omega_0(M) \to \mathcal{M}_*^\phi \otimes_{\mathbf{C}} \omega_\infty(M),$$

which will be called the *connection map*. Before we go further we will need to make the ring $\mathcal{M}_*^\phi$ more explicit.

Remarks 12.12 *The ring $\mathcal{M}_*^\phi$.*

The structure of the ring $\mathcal{M}_*^\phi$ is somewhat complicated. For the study of its structure we have to introduce some theta functions for the modulus q. Let us define the basic theta function θ by the formula

$$\theta(z) = \prod_{n \geq 0}(1 - q^n z^{-1}) \prod_{n > 0}(1 - q^n z).$$

This expression is a holomorphic function on $\mathbf{C}^*$ and its divisor is $\sum_{n \in \mathbf{Z}}[q^n]$. Put $\theta_x(z) = \theta(x^{-1}z)$ for any $x \in \mathbf{C}^*$. The divisor of θ_x is $\sum_{n \in \mathbf{Z}}[xq^n]$. The fundamental formula is

$$\theta_x(qz) = \frac{x}{-qz}\theta_x(z).$$

The function $e(c)_* := \frac{\theta_c}{\theta}$ has the property $e(c)_*(qz) = c\,e(c)_*(z)$. Therefore $e(c)_*^{-1}e(c)$ is a ϕ-invariant element of $\mathcal{M}_*$. It is not difficult to show that $\mathcal{M}_*^\phi$ is equal to the ring $\mathcal{M}(E)[\{\frac{e(c)}{e(c)_*}\}]$. We note that $e(1)_* = 1$, $e(qc)_* = ze(c)_*$ and that in general $e(c_1)_* e(c_2)_* \neq e(c_1 c_2)_*$. Therefore $\mathcal{M}(E)[\{\frac{e(c)}{e(c)_*}\}]$ is a "twisted group algebra" of the group $\mathbf{C}^*$ over the field $\mathcal{M}(E)$.

Lemma 12.13 *The ring $\mathcal{M}_*^\phi$ has no zero divisors.*

Proof: Let Λ be a finitely generated subgroup of $\mathbf{C}^*$. Then Λ is a product $\mu_n \times \Lambda_0$. The group Λ_0 is free of rank m and μ_n is the group of the n-th roots of unity. The subring $\mathcal{M}(E)[\frac{e(c)}{e(c)_*}]_{c \in \Lambda}$ of $\mathcal{M}_*^\phi$ can be written as $\mathcal{M}(E)[B, X_1, X_1^{-1}, \ldots, X_m, X_m^{-1}]$ with $B = \frac{e(\zeta_n)}{e(\zeta_n)_*}$. The only relation is $B^n = (\frac{\theta_{\zeta_n}}{\theta})^{-n} \in \mathcal{M}(E)$. Since $(\frac{\theta_{\zeta_n}}{\theta})^{-m} \notin \mathcal{M}(E)$ for $1 \leq m < n$, one sees that the equation $Y^n - (\frac{\theta_{\zeta_n}}{\theta})^{-n}$ is irreducible over $\mathcal{M}(E)$. This implies that $\mathcal{M}(E)[B]$ is a field and so $\mathcal{M}(E)[\frac{e(c)}{e(c)_*}]_{c \in \Lambda}$ has no zero divisors. ∎

This makes the ring $\mathcal{M}_*^\phi$ in some sense more pleasant. It is of course not possible to give an ϕ-equivariant embedding of $\mathcal{M}_*^\phi$ in $\mathcal{M}(\mathbf{C}^*)$. This is a justification for our use of symbols.

We shall now give an interpretation of the ring $\mathcal{M}_*^\phi$ as ring of meromorphic functions on a natural space. Several copies of $\mathbf{C}$ and $\mathbf{C}^*$ have to be introduced. In order to distinguish them we will add a variable to the space. One considers first the sequence of Riemann surfaces $\mathbf{C}_t \to \mathbf{C}_z^* \to E$ with $z = e^{2\pi i \tau t}$. Further coverings $\mathbf{C}_t \to \mathbf{C}_u^*$ and $\mathbf{C}_v \to \mathbf{C}_u^*$ are defined by the formulas $u = e^{2\pi i t}$ and $u = e^{2\pi i v}$. The action of ϕ on the various spaces is indicated by the action of ϕ on the variables. The actions are given by $\phi(t) = t + 1$, $\phi(u) = u$ and $\phi(v) = v$. The difference ring $\mathcal{M}_* = \mathcal{M}(\mathbf{C}_z^*)[e(c), l]_{c \in \mathbf{C}^*}$ is mapped to $\mathcal{M}(\mathbf{C}_t)[e(c)]_{c \in \mu_\infty}$ in the following way. Choose a $\mathbf{Q}$-linear subspace L of $\mathbf{C}$ with $L \oplus \mathbf{Q} = \mathbf{C}$. Every element $c \in \mathbf{C}^*$ is then written as $c_0 c_1$ with $c_0 \in \mu_\infty$ and $c_1 = e^{2\pi i a}$ with $a \in L$. The map is given by the usual inclusion $\mathcal{M}(\mathbf{C}_z^*) \subset \mathcal{M}(\mathbf{C}_t)$ and $l \mapsto t$ and $e(c) \mapsto e(c_0) e^{2\pi i a t}$. It can be verified that $\mathcal{M}_* \to \mathcal{M}(\mathbf{C}_t)[e(c)]_{c \in \mu_\infty}$ is a ϕ-equivariant and injective homomorphism. The ring $\mathcal{M}(\mathbf{C}_t)[e(c)]_{c \in \mu_\infty}$ can also be written as $\mathcal{M}(\mathbf{C}_t)[f(\lambda)]_{\lambda \in \mathbf{Q}}$ with $f(\lambda) := e^{-2\pi i \lambda t} e(e^{2\pi i \lambda})$. The $f(\lambda)$'s are ϕ-invariant elements. They have the relations $f(\lambda_1) f(\lambda_2) = f(\lambda_1 + \lambda_2)$ and $f(\lambda) = u^{-\lambda}$ if $\lambda \in \mathbf{Z}$. The ring of ϕ-invariant elements of $\mathcal{M}(\mathbf{C}_t)[e(c)]_{c \in \mu_\infty}$ is therefore $\mathcal{M}(\mathbf{C}_u^*)[f(\lambda)]_{\lambda \in \mathbf{Q}}$. In particular, $f(1/m)^m = f(1) = u^{-1}$. Now one uses the map $\mathbf{C}_v \to \mathbf{C}_u^*$. This offers an embedding, which is of course ϕ-equivariant, of $\mathcal{M}(\mathbf{C}_u^*)[f(\lambda)]_{\lambda \in \mathbf{Q}}$ into $\mathcal{M}(\mathbf{C}_v)$ given by the usual inclusion $\mathcal{M}(\mathbf{C}_u^*) \subset \mathcal{M}(\mathbf{C}_v)$ and $f(\lambda) \mapsto e^{-2\pi i \lambda v}$. The final conclusion is:

$\mathcal{M}_*^\phi$ can be embedded into $\mathcal{M}(\mathbf{C}_v)$. More precisely $\mathcal{M}_*^\phi$ embeds into the field $\cup_{m \geq 1} \mathcal{M}(\mathbf{C}_u^*)(e^{2\pi i v/m})$. The connection map has coordinates in this field.

With this terminology one is able to understand the work of G.D. Birkhoff on q-difference equations, in particular, the papers [10] and [13].

In the first paper, the function $H := q^{(t^2-t)/2} = e^{\pi i \tau (t^2-t)}$ is introduced as a solution of $\phi(H) = zH$. The function $z^b = e^{2\pi i \tau bt}$ is introduced as a solution of the equation $\phi(f) = e^{2\pi i \tau b} f$. Here is of course a slight problem! The q-difference equations studied in the first paper are essentially regular singular. Birkhoff asserts that the connection matrix is an invertible meromorphic matrix on $\mathbf{C}_t$ with period 1. That would mean a matrix with coordinates in $\mathcal{M}(\mathbf{C}_u^*)$. This is true if the roots of unity do not appear in the formal classification of the q-difference equation. In general however this is wrong for the canonical choice of the connection matrix. In the two papers, the formal classification of the equation is given by producing a fundamental matrix with coordinates in $\mathcal{M}(\mathbf{C}_t)$. Here again the use of the expression z^b gives rise to some confusion. ∎

We consider now the category $\mathbf{G_{rs}}$-**Triples**. An object of this category is a triple (V_0, V_∞, T) where V_0 and V_∞ are finite dimensional $\mathbf{G_{rs}}$-representations of the same dimension and with

$$T : \mathcal{M}_*^\phi \otimes_{\mathbf{C}} V_0 \to \mathcal{M}_*^\phi \otimes_{\mathbf{C}} V_\infty$$

a $\mathcal{M}_*^\phi$-linear and $\mathbf{G_{rs}}$-equivariant isomorphism. The category $\mathbf{G_{rs}}$-**Triples** is in an obvious way a neutral Tannakian category with fibre functor $(V_0, V_\infty, T) \mapsto V_0$.

Theorem 12.14

> (a) *The functor $M \mapsto (\omega_0(M), \omega_\infty(M), S_M)$ from the category of the regular singular q-difference modules over $\mathbf{C}(z)$ to the category $\mathbf{G_{rs}}$-Triples is an equivalence of $\mathbf{C}$-linear tensor categories.*

> (b) *The ring $\mathcal{M}_*^\phi$ is considered, as above, as a subring of the field of meromorphic functions $\mathcal{M}(\mathbf{C}_v)$. The difference Galois group G, seen as an algebraic subgroup of $Aut(\omega_0(M))$, is the smallest algebraic subgroup which contains the image of $\mathbf{G_{rs}}$ and the $S_M^{-1}(a) S_M(b)$ for all $a, b \in \mathbf{C}_v$ such that $S_M(a)$ and $S_M(b)$ are invertible matrices.*

Proof: (a) The proof is rather similar to that of 12.9. We will only give the proof that any object (V_0, V_∞, T) of $\mathbf{G_{rs}}$-**Triples** is isomorphic to some $(\omega_0(M), \omega_\infty(M), S_M)$.
The module $\mathcal{M}(\mathbf{C})[\{e(c)\}, l] \otimes_{\mathbf{C}} V_0$ has an action of ϕ and $\mathbf{G_{rs}}$. The two actions commute. Let M_0 denote the set of invariants under the action of $\mathbf{G_{rs}}$. Then M_0 is a vector space over $\mathcal{M}(\mathbf{C})$ with a ϕ-action. Moreover the canonical map

$$\mathcal{M}(\mathbf{C})[\{e(c)\}, l] \otimes_{\mathcal{M}(\mathbf{C})} M_0 \to \mathcal{M}(\mathbf{C})[\{e(c)\}, l] \otimes_{\mathbf{C}} V_0$$

is an isomorphism, which commutes with the actions of ϕ and $\mathbf{G_{rs}}$. One defines M_∞ in a similar way. The $\mathbf{G_{rs}}$ and ϕ-equivariant map T induces, by taking $\mathbf{G_{rs}}$-invariants, a ϕ-equivariant isomorphism

$$\mathcal{M}(\mathbf{C}^*) \otimes M_0 \to \mathcal{M}(\mathbf{C}^*) \otimes M_\infty.$$

As in 12.3.1, the modules M_0 and M_∞ glue to a q-difference module M over $\mathbf{C}(z)$. One easily verifies that the triple of M is isomorphic to the given triple (V_0, V_∞, T).

(b) It is not difficult to show (along the lines of the proof of (b) of Theorem 12.9) that the linear algebraic subgroup of $\mathrm{Aut}(V_0)$ corresponding to an object (V_0, V_∞, T) of $\mathbf{G_{rs}}$-**Triples** is the smallest algebraic subgroup containing the image of $\mathbf{G_{rs}}$ and the $T(a)^{-1}T(b)$ (for all $a, b \in \mathbf{C}_v$ such that $T(a)$ and $T(b)$ are invertible matrices). ∎

12.3.4   Inverse problems

We will make the last theorem more explicit by considering a subclass of the regular singular q-difference equations over $\mathbf{C}(z)$. A q-difference module M over $\mathbf{C}(z)$ will be called *semi-regular of type n* if

(a) $k_0 \otimes M$ is isomorphic to a direct sum of one dimensional modules $k_0 e_j$ with the property $\Phi(e_j) = \lambda_j e_j$ and $\lambda_j^n \in \{q^m | m \in \mathbf{Z}\}$.

(b) A similar condition for $k_\infty \otimes M$.

The following statements are easily seen to be equivalent:

(1) M is semi-regular of type n and of dimension d over $\mathbf{C}(z)$.

(2) The vector spaces

$$\tau_0(M) := ker(\Phi - 1, k_0[e(e^{2\pi i/n}), e(e^{2\pi i\tau/n})] \otimes M) \text{ and}$$

$$\tau_\infty(M) := ker(\Phi - 1, k_\infty[e(e^{2\pi i/n}), e(e^{2\pi i\tau/n})] \otimes M)$$

have dimension d over $\mathbf{C}$.

(3) Let $y(qz) = Ay(z)$ denote a matrix equation corresponding to M, then the equation is both over k_0 and k_∞ equivalent to a matrix equation with a diagonal matrix with entries in the group $\{\lambda \in \mathbf{C}^* | \lambda^n \in q^{\mathbf{Z}}\}$.

We also introduce $\tau_*(M)$ for a semi-regular q-difference module M by

$$\tau_*(M) = ker(\Phi - 1, \mathcal{M}(\mathbf{C}^*)[e(e^{2\pi i/n}), e(e^{2\pi i\tau/n})] \otimes M).$$

Let H denote the group of the automorphisms of $\mathbf{C}(z)[e(e^{2\pi i/n}), e(e^{2\pi i\tau/n})]$ over $\mathbf{C}(z)$ which commute with the action of ϕ. Then H is the product of two cyclic groups of order n. We need to investigate the set K_n of the ϕ-invariants of the ring $\mathcal{M}(\mathbf{C}^*)[e(e^{2\pi i/n}), e(e^{2\pi i\tau/n})]$. As in 12.12 one can verify that

$$K_n = \mathcal{M}(E)\Big[\frac{e(e^{2\pi i/n})}{e(e^{2\pi i/n})_*}, \frac{e(e^{2\pi i\tau/n})}{e(e^{2\pi i\tau/n})_*}\Big].$$

Let us denote the elements $\frac{e(e^{2\pi i/n})}{e(e^{2\pi i/n})_*}$ and $\frac{e(e^{2\pi i\tau/n})}{e(e^{2\pi i\tau/n})_*}$ by E_1 and E_2. They satisfy the equations $E_1^n = f_1$ and $E_2^n = f_2$, where f_1, f_2 are certain elements of $\mathcal{M}(E)$. It follows that K_n is a finite field extension of $\mathcal{M}(E)$ of degree at most n^2. The group H acts on K_n. Put $\zeta_n = e^{2\pi i/n}$. For every tuple (j_1, j_2) with $0 \leq j_1, j_2 < n$, there is a unique element $\sigma \in H$ such that $\sigma e(e^{2\pi i/n}) = \zeta_n^{j_1} e(e^{2\pi i/n})$ and $\sigma e(e^{2\pi i\tau/n}) = \zeta_n^{j_2} e(e^{2\pi i\tau/n})$. For this σ one has $\sigma(E_i) = \zeta_n^{j_i} E_i$ for $i = 1, 2$. Hence K_n is a Galois extension of $\mathcal{M}(E)$ with Galois group H.

One can verify that K_n is actually the function field of an elliptic curve F over $\mathbf{C}$. The group H acts on F as the group of the translations over the group $F[n]$ of the points of order n of the elliptic curve F. The curve E is then isomorphic to the quotient $F/F[n]$. Thus the curve F is actually isomorphic with E.

We introduce the neutral tensor category **H-Triples** analogous with the notations before, as the category having objects (V_0, V_∞, T) with:
(1) V_0, V_∞ are finite dimensional vector spaces over $\mathbf{C}$ provided with an H-action. The two vector spaces are supposed to have the same dimension.
(2) $T : K_n \otimes V_0 \to K_n \otimes V_\infty$ is a K_n-linear and H-equivariant isomorphism.

For a semi-regular q-difference module M of type n there are natural K_n-linear and H-equivariant isomorphisms $K_n \otimes \tau_0(M) \to \tau_*(M)$ and $K_n \otimes \tau_\infty(M) \to \tau_*(M)$. The resulting "connection map"

$$K_n \otimes \tau_0(M) \to K_n \otimes \tau_\infty(M)$$

will again be denoted by S_M.

Corollary 12.15

1. *The functor $M \mapsto (\tau_0(M), \tau_\infty(M), S_M)$ from the category of the semi-regular q-difference modules of type n over $\mathbf{C}(z)$ to the category* **H-Triples** *is an equivalence of $\mathbf{C}$-linear tensor categories.*

2. *The difference Galois group of a semi-regular q-difference module of degree n over $\mathbf{C}(z)$ is the smallest algebraic subgroup G of $Aut(\tau_0(M))$ such that the image of H belongs to G and $S_M(a)^{-1}S_M(b) \in G$ for all $a, b \in F$ such that $S_M(a)$ and $S_M(b)$ are invertible maps.*

Let us study **H**-triples more closely. The *Tannakian group* associated with the triple is the group described in the corollary above. The main question is to find out what the possibilities for this Tannakian group are.

We make a slight extension of the notion of **H**-triples by introducing **H, K, L**-triples. The K, L are fields, chosen once for all, such that $\mathbf{C} \subset K \subset L$. The field L is supposed to be a Galois extension of K with a Galois group H which is a finite commutative group.

A triple (V_0, V_∞, T) is called a **H, K, L**-triple if $\rho_0 : H \to \mathrm{Aut}(V_0)$ and $\rho_\infty : H \to \mathrm{Aut}(V_\infty)$ are two representations of the same finite dimension; $T : L \otimes V_0 \to L \otimes V_\infty$ is an L-linear and H-equivariant isomorphism. The Tannakian group of a triple can be seen to be the smallest algebraic subgroup G of $\mathrm{Aut}(V_0)$ such that:

(a) G contains the image of H.

(b) For one (or any) linear isomorphism $A : V_\infty \to V_0$ the map AT lies in a coset $BG(L)$ with $B \in \mathrm{Aut}(V_0)$.

Indeed, T is already defined over a finitely generated and H-invariant **C**-subalgebra O_L of L. Then it is easily seen that G is the smallest algebraic subgroup of $\mathrm{Aut}(V_0)$ which contains the image of H and $T(a)^{-1}T(b)$ for all maximal ideals a, b of O_L. The last statement translates easily into (a) and (b).

Let, for the moment, Z denote the smallest algebraic subgroup of $\mathrm{Aut}(V_0)$ such that condition (b) holds (and thus Z is the smallest algebraic group containing all $T(a)^{-1}T(b)$ with the notation above). Then Z is connected and thus $Z \subset G^o$. The H-equivariance of T implies that $\rho_0(h)^{-1}Z\rho_0(h) = Z$ for every $h \in H$. Then Z has finite index in the group Z' generated by Z and $\rho_0(H)$. Thus Z' is an algebraic group. It is clear that Z' satisfies (a) and (b). Hence $Z' = G$ and $Z = G^o$.

A necessary condition for an algebraic subgroup G of $\mathrm{Aut}(V_0)$ to be the Tannakian group of a triple is therefore:

There is a group homomorphism $\rho : H \to G$ such that $H \to G \to G/G^o$ is surjective.

The problem is to show that this condition is also sufficient. At the moment we cannot answer this question. The following weaker result holds.

Theorem 12.16 *Suppose that $K/\mathbf{C}$ has transcendence degree one. Let $\rho : H \to \mathrm{Aut}(V)$ be a finite dimensional complex representation. Let G be an algebraic subgroup of $\mathrm{Aut}(V)$ such that:*
(1) $\rho(H) \subset G$ and $\rho : H \to G \to G/G^o$ is surjective.

(2) $\rho(H)$ *lies in* $N(G^o)^o$, *i.e. the component containing the neutral element of the normalizer* $N(G^o)$ *of* G^o *in* $Aut(V)$.
Then G *is the Tannakian group of a* **H, K, L**-*triple.*

Proof: We start by proving that any finite commutative subgroup A of a connected algebraic group $Z \subset Aut(V)$ lies in a (maximal) torus of Z. (This is probably well known, but we could not find a reference for this). It suffices to show that A lies in a Borel subgroup of Z ([29], 19.4). Let A be generated by s elements $a_1, \ldots, a_s$. Let B denote a Borel subgroup of Z containing a_1 and let X denote the projective variety Z/B. The element a_1 acts on X by right multiplication and the set Y of fixed points of this action corresponds to the set of Borel subgroups containing a_1. Furthermore, Y is a nonempty closed subset of X. By induction there is a torus S containing $a_2, \ldots, a_s$. Since Y is a projective variety and S is connected and solvable, there is a fixed point for the action of S on Y. This fixed point corresponds to a Borel subgroup containing A.

The triple will be (V, V, T) where $T \in G^o(L) \subset Aut(L \otimes V)$ has still to be specified. The group $N(G^o)^o$ is the component, containing 1, of the normalizer $N(G^o)$ of G^0 in $Aut(V)$. Let $S \subset N(G^o)^o$ be a minimal torus which contains $\rho(H)$. The rank of S is equal to the minimal number of generators of $\rho(H)$. Let X denote the group of characters of S. One decomposes V as $\oplus V_{\chi \in X}$, such that for all $s \in S$ and all χ, the element s acts on V_χ as multiplication by $\chi(s)$.

The dual of the group H is written as $\hat{H}$. There are elements $e_d \in L$, $d \in \hat{H}$ such that $L = \oplus_{d \in \hat{H}} K e_d$ and $h(e_d) = d(h)e_d$ for all $h \in H$ and all $d \in \hat{H}$. The homomorphism $\rho : H \to S$ induces a homomorphism $\hat{\rho} : X \to \hat{H}$. We consider $E \in Aut(L \otimes V)$ given by: the restriction of E with respect to each V_χ is the multiplication by $e_{\hat{\rho}\chi}$. The element E belongs to $S(L)$ and hence also to the normalizer of $G^o(L)$.

The Galois group H of L/K acts in a natural way on the elements $A \in Aut(L \otimes V)$. The image of A under $h \in H$ is written as ${}^hA$. By construction ${}^hE = \rho(h)E = E\rho(h)$ for every $h \in H$. An element $T \in G^o(L)$ is H-equivariant if and only if $T(h_L \otimes \rho(h)) = (h_L \otimes \rho(h))T$ for all $h \in H$. (Here h_L denotes the ordinary action of h on L). The last condition can be written as ${}^hT = \rho(h)^{-1}T\rho(h)$ for all $h \in H$. This leads to the following translation of the H-equivariance of elements of $G^o(L)$:

$T \in G^o(L)$ *is* H-*equivariant if and only if* $T = E^{-1}SE$ *with* $S \in G^o(K)$.

We want to find an $S \in G^o(K)$ such that $T := E^{-1}SE$ does not lie in $BZ(L)$ for any proper algebraic subgroup Z of G^o and any $B \in Aut(V)$. There are elements $g_1, \ldots, g_s \in G^o$ which generate G^o as an algebraic subgroup of

$Aut(V)$. Choose a transcendental element $t \in K$, the subalgebra $\mathbf{C}[t]$ of K and its integral closure O_L in L. Take distinct $a_0, a_1, \ldots, a_s \in \mathbf{C}$. Above each maximal ideal $(t - a_j)$ of $\mathbf{C}[t]$ we choose a maximal ideal b_j of O_L. We may (and do) choose the a_j such that E is defined and invertible at each b_j. The element $S \in G^o(K)$ will be chosen in $G^o(\mathbf{C}(t))$. Hence S is seen as a rational map of $\mathbf{P}^1 \to G^o$. This rational map is chosen such that $S(a_0) = 1$ and $S(a_j) = E(b_j) g_j E(b_j)^{-1}$ for $j = 1, \ldots, s$. Such an S exists since G^o is a rational variety. The element S has coefficients in a localization $\mathbf{C}[t]_f$ of $\mathbf{C}[t]$. Then $T = E^{-1}SE$ has coefficients in some localization $(O_L)_g$. The b_j are still maximal ideals of $(O_L)_g$. By construction $T(b_0) = 1$ and $T(b_j) = g_j$ for $j = 1, \ldots, s$. The set $\{T(b_0)^{-1}T(b_j) | j = 1, \ldots, s\}$ generates G^o as an algebraic group. This shows that T does not lie in any $BZ(L)$, with Z a proper subgroup of G^o and $B \in Aut(V)$. ∎

The next corollary follows from 12.15 and 12.16 for the choice $K = \mathcal{M}(E)$, $L = \mathcal{M}(F)$ (for suitable F) and $H \to Z$ a surjective homomorphism.

Corollary 12.17 *A linear algebraic group $G \subset Gl(n, \mathbf{C})$ is the difference Galois group of a regular singular q-difference equation over $\mathbf{C}(z)$ if*
(1) G contains a finite commutative subgroup Z which has at most two generators and which is mapped surjectively to G/G^0.
(2) Z lies in the connected component of the normalizer of G^o in $Gl(n, \mathbf{C})$.

We *conjecture* that in this corollary condition (2) is superfluous.
As an example we will work out the case where the group G^o is a torus.

Example 12.18 *G with G^o a torus*

The method used in this example is a variation on the proof of lemma 8.12. The character group of the torus G^o is written as X. There is an H-action on X. We claim that it is enough to produce a homomorphism $T : X \to L^*$ such that:

(a) T is H-equivariant.

(b) The preimage $T^{-1}(\mathbf{C}^*)$ is 0.

Let us assume that we have such a T. We may identify G^o with the group of diagonal elements of some $Gl(n, \mathbf{C})$. Let χ_i be the character of G^0 that corresponds to the projection onto the ith diagonal element. Finally, let $S = diag(T(\chi_1), \ldots, T(\chi_n))$. Clearly $S \in G^o(L)$. An easy calculation shows that conditions (a) and (b) imply that $B \in G^o(\mathbf{C})$ and that S is H-equivariant and that S is not contained in some $BZ(L)$ with Z a proper algebraic subgroup of G^o.

Let $\mathbf{Z}[H]$ denote the group ring of H over $\mathbf{Z}$. This is an example of a free $\mathbf{Z}$-module with an H-action. In fact any H-module X (free and finitely generated over **Z**) can be embedded in $\mathbf{Z}[H]^N$ for some $N \geq 1$. It suffices to produce an element T for the last module having the required properties.

To do this, we will find elements $f_1, \ldots, f_N \in L^*$ such that $\prod_{h \in H, 1 \leq j \leq N} h(f_j)^{m_{h,j}} \in \mathbf{C}^*$ for any integers $m_{h,j}$, implies that all $m_{h,j}$ are 0. Defining $T(\sum_{h \in H} m_{h,1}h, \ldots, \sum_{h \in H} m_{h,N}h) = \prod_{h \in H, 1 \leq j \leq N} h(f_j)^{m_{h,j}}$ then gives the desired T. We assume, as in the Theorem 12.16 that K has transcendence degree one over **C**. The fields K and L are seen as the function fields of algebraic curves F_1, F_2. There is given a Galois covering $\pi : F_2 \to F_1$ with Galois group H.

We will construct the f_i, one by one. Let N denote the order of H. Take an unramified point $a \in F_1$. Then $\pi^{-1}a = \{a_1, \ldots, a_N\}$. There is a rational function f_1 on F_2 such that $f_1(a_1) = 0$ and $f_1(a_2) = \ldots = f_1(a_N) = 1$. Suppose that $F := \prod_{h \in H} h(f_1)^{m_h} \in \mathbf{C}^*$. If some $m_h \neq 0$ then F has a pole or a zero in some point a_j. Hence all m_h are 0.

For the construction of f_2 we take an unramified point $b \in F_2$ such that f_1 has no zeroes or poles on the set $\pi^{-1} = \{b_1, \ldots, b_N\}$. One takes for f_2 a rational function on F_2 such that f_2 has non zero values on $\pi^{-1}a$ and $f_2(b_1) = 0$ and $f_2(b_1) = \ldots = f_2(b_N) = 1$. Consider a product $F := \prod_h h(f_1)^{m_{h,1}} \prod_h h(f_2)^{m_{h,2}}$. If one of the exponents $m_{h,i}$ is not zero then F has either a pole or a zero in the set $\pi^{-1}(a) \cup \pi^{-1}(b)$. So the pair f_1, f_2 has already the required property. By induction one can construct a tuple $(f_1, \ldots, f_n)$ with the required property for any n. This finishes the verification of the conjecture above for tori.

Example 12.19 *Order one equations over* $\mathbf{C}(z)$

Consider the order one equation $y(qz) = a(z)y(z)$ for *any* $a \in \mathbf{C}(z)^*$. The divisor of a is $m_0[0] - m_\infty[\infty] + \sum_{a_j \in \mathbf{C}^*} m_j[a_j]$. Write $a = (-z)^{m_0} c_0 b_0$ with c_0 a constant and b_0 with $b_0(0) = 1$. Similarly $a = (-z)^{m_\infty} c_\infty b_\infty$ with c_∞ a constant and b_∞ with $b_\infty(\infty) = 1$. The expression $f_0 := \prod_{j \geq 0} b_0(q^j z)^{-1}$ converges on **C** and has the property $f_0(qz)^{-1} a f_0(z) = (-z)^{m_0} c_0$. The function $H(z) := \frac{1}{z\theta(z)}$ satisfies $H(qz) = -zH(z)$. We find a solution $f_0 H^{m_0} e(c_0) \in \mathcal{M}_*$ of our equation. Put $f_\infty := \prod_{j \geq 1} b_\infty(q^{-j} z)$. This function converges on $\mathbf{C} \cup \{\infty\}$ and gives a second solution $f_\infty H^{m_\infty} e(c_\infty) \in \mathcal{M}_*$. The connection matrix, in an extended sense because we have allowed singularities of a at 0 and ∞, is equal to

$$f_\infty f_0^{-1} H^{m_\infty - m_0} e(c_\infty c_0^{-1}) \in \mathcal{M}_*^\phi.$$

Let us write $\mathcal{S}(a)$ for the connection matrix of $a \in \mathbf{C}(z)^*$. The map $a \mapsto \mathcal{S}(a)$ is multiplicative (as it should be).

The connection matrix of the equation $y(qz) = (1 - d^{-1}z)y(z)$ can explicitly be calculated and gives the answer $\frac{-d\theta_d}{\theta} e(d^{-1})$. An arbitrary element $a \in \mathbf{C}(z)^*$

can be written in the form $a = c(-z)^m \prod_j (1 - a_j^{-1} z)^{m_j}$ with a_j distinct elements of $\mathbf{C}^*$ and the $m_j \in \mathbf{Z}$. The connection matrix of a is therefore:

$$\prod_j (-a_j)^{m_j} \cdot \prod_j (\frac{\theta_{a_j}}{\theta})^{m_j} \cdot e(\prod_j a_j^{-m_j}).$$

The equation corresponding to a is regular if $m = 0, c = 1, \sum m_j = 0$ and $\prod_j (-a_j)^{m_j} = 1$. In that case $\mathcal{S}(a) = \prod_j (\frac{\theta_{a_j}}{\theta})^{m_j}$ is a meromorphic function on E.

Another interesting equation is $y(qz) = y(z) + a$ with $a \in \mathbf{C}(z)$. This equation is regular if a has a zero at $z = 0$ and $z = \infty$. In that case the "connection matrix" is the convergent sum $\sum_{j \in \mathbf{Z}} a(q^j z)$. For $a = \frac{z}{(1-z)^2}$ the connection matrix has as a function on E a pole of order 2 at $1 \in E$. The connection matrix is therefore a linear combination of the Weierstrass function w and 1. In some texts, this sum $\sum_j \frac{q^j z}{(1-q^j z)^2}$ is taken as a definition of the Weierstrass function. For $a = \frac{z}{(1-z)^3}$ the connection matrix has (as a function on E) a pole of order 3 at $1 \in E$. This connection matrix does not belong to the subfield $\mathbf{C}(w)$ of $\mathcal{M}(E)$.

Example 12.20 *Some equations of order two*

In this example we investigate the possibilities for q-difference equations over $\mathbf{C}(z)$ of order two and with difference Galois group $\mathbf{D}_\infty$, i.e. the automorphisms of $V = \mathbf{C}v_1 + \mathbf{C}v_2$ which have determinant 1 and permute the two lines $\{\mathbf{C}v_1, \mathbf{C}v_2\}$.

Let us study first the *semi-regular modules.*
The group $\mathbf{D}_\infty$ is generated by the torus $\{ \begin{pmatrix} c & 0 \\ 0 & c^{-1} \end{pmatrix} \mid c \in \mathbf{C}^* \}$ and $k :=$
$\begin{pmatrix} 0 & -1 \\ 1 & 0 \end{pmatrix}$, an element of order 4. The semi-regular module will be of type 4 and is given by the following choices:
(1) A homomorphism $\rho : H \to \mathrm{Aut}(V)$ with image the group generated by k. Here, H is the Galois group of $K_4/K = \mathcal{M}(E)$.
(2) A connection map $T = \begin{pmatrix} a & 0 \\ 0 & a^{-1} \end{pmatrix}$.

The equivariance of $a \in K_4^*$ has the explicit meaning: $h(a) = a$ if $\rho(h)$ is an even power of k and $h(a) = a^{-1}$ if $\rho(h)$ is an odd power of k. Let $L \subset K_4$ denote the quadratic extension of K of the elements of K_4, which are fixed under all $h \in H$ such that $\rho(h)$ is an even power of k. The non-trivial automorphism of L/K is denoted by σ. Then $a \in L$ and $a\sigma(a) = 1$. The choice (2) is therefore equivalent to the choice of an a in L with $a\sigma a = 1$ and $a \neq 1, -1$.

Let M be a *general* q-difference module over $\mathbf{C}(z)$ with difference Galois group $\mathbf{D}_\infty$. Let PV denote the Picard-Vessiot ring of the equation. Then the set of invariants of PV under $\mathbf{G}_m$ is a quadratic extension of $\mathbf{C}(z)$, having an extension of the action of ϕ. There are three possibilities for this quadratic extension $\mathbf{C}(z)[t]$, namely $t = e(-1), e(e^{\pi i \tau}), e(-e^{\pi i \tau})$.

In the first case, the $\mathbf{D}_\infty$-torsor associated to M is trivial. This means that M can be represented by a matrix equation

$$y(qz) = \begin{pmatrix} 0 & -A^{-1} \\ A & 0 \end{pmatrix} y(z),$$

with $A \in \mathbf{C}(z)$.

On the other hand, a difference equation of the type above has a difference Galois group G contained in $\mathbf{D}_\infty$. A criterion on $A \in \mathbf{C}(z)^*$ for $G = \mathbf{D}_\infty$ is the following:

Proposition 12.21 *The above difference equation has Galois group $\mathbf{D}_\infty$ if and only if the equation $y(q^2 z) = (-A(z)A(qz)^{-1})^N y(z)$ has only for $N = 0$ a non trivial solution in $\mathbf{C}(z)$.*

Proof: Let $(y_1, y_2) \neq 0$ denote a solution of the matrix equation with y_1, y_2 in the Picard-Vessiot ring PV of the equation. Then $\phi^2(y_1) = -A(z)A(qz)^{-1}y_1(z)$. Further $\phi(y_1 y_2) = -y_1 y_2$ and so $e(-1) \in PV$. Hence PV contains two independent solutions y_1 and $e(-1)y_1$ of the equation $\phi^2(y) = -A(z)A(qz)^{-1}y$. A fundamental matrix for the matrix equation is $\begin{pmatrix} y_1 & e(-1)y_1 \\ y_2 & -e(-1)y_2 \end{pmatrix}$, with $y_2 = A(q^{-1}z)\phi^{-1}y_1$. This proves that the Picard-Vessiot ring of the order two equation $\phi^2(y) = -A(z)A(qz)^{-1}y$ is equal to PV.
The equation $\phi^2(y) = -A(z)A(qz)^{-1}y$ is seen as a difference equation with parameter q^2 instead of q. The ring PV seen as a difference ring with respect to ϕ^2 contains a non zero solution of the equation $\phi^2(y) = -A(z)A(qz)^{-1}y$. But PV is not the Picard-Vessiot ring PV' of the equation since PV still has the non trivial ϕ^2-invariant ideal $(e(-1) - 1)$. It can be seen that this ideal is maximal among the set of ϕ^2-invariant ideals. Hence PV' can be identified with $PV/(e(-1) - 1)$. The condition of the criterion is equivalent to $PV' \cong \mathbf{C}(z)[Y, Y^{-1}]$, with $\phi^2(Y) = -A(z)A(qz)^{-1}Y$ and no relations for Y. Using $PV' = PV/(e(-1) - 1)$, one finds that the criterion is also equivalent to $G = \mathbf{D}_\infty$. ∎

In the second case, the module $\mathbf{C}(t) \otimes M$ decomposes as a direct sum of two conjugated difference modules $\mathbf{C}(t)e_1$ and $\mathbf{C}(t)e_2$ with ϕ-actions given by $\phi(e_1) = (a_0 + a_1 t)e_1$ and $\phi(e_2) = (a_0 - a_1 t)e_2$, with $a_0, a_1 \in \mathbf{C}(z)$ and $a_1 \neq 0$.

The module M is identified with $\mathbf{C}(z)(e_1 + e_2) + \mathbf{C}(z)t(e_1 - e_2)$. The second exterior power $\Lambda^2 M$ has basis $te_1 \wedge e_2$ over $\mathbf{C}(z)$ and is trivial. It follows that $q^{1/2}(a_0 + a_1 t)(a_0 - a_1 t) = \phi(f)f^{-1}$ for some $f \in \mathbf{C}(z)$. The term $a_0 + a_1 t$ may be changed into $(a_0 + a_1 t)\phi(g)g^{-1}$ with $g \in \mathbf{C}(t)^*$. Using this one can normalize the choice of $a_0 + a_1 t$ such that $a_0^2 - za_1^2 = q^{-1/2}$. The module M is generated over $\mathbf{C}(z)$ by the elements $e_1 + e_2$ an $t(e_1 - e_2)$. This leads to a matrix equation for M with respect to this basis of the form

$$y(qz) = \begin{pmatrix} a_0 & q^{1/2}za_1 \\ a_1 & q^{1/2}a_0 \end{pmatrix} y(z).$$

The matrix equation above, with $a_0, a_1 \in \mathbf{C}(z)$, $a_0 \neq 0$ and $q^{1/2}(a_0^2 - za_1^2) = 1$, can be seen as the standard form of a q-difference equation over $\mathbf{C}(z)$ with group $\mathbf{D}_\infty$ and with a Picard-Vessiot ring containing $e(e^{\pi i \tau})$.

In the third case there is a similar standard form with $q^{1/2}$ replaced by $-q^{1/2}$.

Remarks

In [27] algorithms for order two q-difference equations are given. In particular the Galois groups of the q-hypergeometric difference equations are determined. There is a theory of q-difference equations for q a root of unity. This theory is developed in [27]. The results are rather different from the case $0 < |q| < 1$ studied in this chapter.

Bibliography

[1] Abramov,S., Paule, P., Petkovšek, M., *q-Hypergeometric Solutions of q-Difference Equations*, manuscript, 1995.

[2] Abramov, S., Bronstein, M., Petkovšek, M., *On Polynomial Solutions of Linear Operator Equations*, manuscript, 1995.

[3] Atiyah, M. and MacDonald, I.G., **Introduction to Commutative Algebra**, Addison-Wesley, Reading, Mass., 1969.

[4] Balser, W., Jurkat, W.B. and Lutz, D.A.,*A General Theory of Invariants for Meromorphic Differential Equations; Part I, Formal Invariants*, Funk. Ekvacioj, bf 22 (1979), 197-221

[5] Batchelder, P.M., **An introduction to Linear Difference Equations** Dover Pub, New York, 1967

[6] Benzaghou, B., *Algèbres de Hadamard*, Bull. Soc. Math. France, **98** (1970), 209-252.

[7] Benzaghou,B. and Bezivin, J.-P., *Propriétés algébriques de suites différentiellement finies*, Bull. Soc. Math. France, **120** (1992), 327-346.

[8] Bialynicki-Birula, A., *On Galois theory of fields with operators*, Amer. J. Math. **84**, (1962), 89-109.

[9] Birkhoff, G.D., *General theory of linear difference equations*, Trans. Amer. Math Soc. **12**, (1911), 243-284.

[10] Birkhoff, G.D., *The generalized Riemann problem for linear differential equations and the allied problem for linear difference and q-difference equations*, Proc. Nat. Acad. Sci. **49**, (1913), 521-568.

[11] Birkhoff, G.D., *Theorem concerning the singular points of ordinary linear differential equations*, Proc. Nat. Acad. Sci. **1**, (1915), 578-581.

[12] Birkhoff, G.D., *Note on linear difference and differential equations*, Proc. Nat. Acad. Sci. **27**, (1941), 65-67.

[13] Birkhoff, G.D., *Note on a canonical form for the linear q-difference system*, Proc. Nat. Acad. Sci. **27**. (1941). 218-222.

[14] Birkhoff, G.D. and Tritzinsky,W.J., *Analytic theory of singular difference equations*, Acta Math., **60**, (1933), 1-89.

[15] Bourbaki, N., **Algèbre**, Chap. 8, "Modules et Anneaux Semi-Simple", Hermann, Paris, 1958.

[16] Braaksma, B.L.J. and Faber, B.F., *Multisummability for some classes of difference equations*, Preprint University of Groningen. May 1995.

[17] Chase, S.U., Harrison, D.K., Rosenberg. A.. *Galois theory and cohomology of commutative rings*, Memoirs of the AMS, **52**. American Mathematical Society, Providence, 1965.

[18] Coddington, E. and Levinson, N., **Theory of Ordinary Differential Equations**, McGraw-Hill, New-York, 1955.

[19] Cohen, R., **Difference Algebra**, Tracts in Mathematics, Number 17, Interscience Press, New York, New York, 1965.

[20] Deligne, P. and Milne, J.S., *Tannakian categories*, Lecture Notes in Mathematics, **900**, 101-228.

[21] Duval, A., *Lemmes de Hensel et factorisations formelle pour les opérateurs aux différences*, Funk. Ekv., **26** (1983), 349-368.

[22] Etingof, P.I., *Galois groups and connection matrices of q-difference equations*. Electronic Research Announcements of the A.M.S., Volume 1, Issue 1, (1995).

[23] Fahim, A., *Extensions Galoisiennes d'algèbres différentielles*, C.R. Acad. Sci. Paris, t. 314, Série I (1992), 1-4.

[24] Foster, O., **Lectures on Riemann Surfaces**, Graduate Texts in Mathematics, **81**, Springer-Verlag, Berlin, Heidelberg, 1981.

[25] Franke, C.H., *Picard-Vessiot theory of linear homogeneous difference equations*, Trans. Am. Math. Soc., **108** (1963), 491-515.

[26] Hendriks, P.A., *An algorithm for determining the Galois group of second order linear difference equations*, to appear in the *J. of Symb. Comp.*.

[27] Hendriks, P.A., *Algebraic aspects of linear differential and difference equations*, thesis University of Groningen, november 1996.

[28] Hendriks, P.A., van der Put, M., *Galois action on solutions of differential equations*, J. Symb. Comp., **14** (1995), 559-576.

[29] Humphreys, J. E., **Linear Algebraic Groups**, Second Edition, Springer-Verlag, New York, 1981.

[30] Immink, G.K., *Asymptotics of Analytic Difference Equations*, Lecture Notes in Math. **1085**, Springer Verlag, Berlin, Heidelberg, New-York, Tokyo, 1984.

[31] Immink, G.K., *On meromorphic equivalence of linear difference operators*, Ann. Inst. Fourier, **40** (1990), 683-699.

[32] Immink, G.K., *Reduction to canonical forms and the Stokes phenomenon in the theory of linear difference equations*, Siam J. Math. Analysis, **22** (1991), 238-259.

[33] Kaplansky, I., **An Introduction to Differential Algebra**, Second Edition, Hermann, Paris, 1976.

[34] Kovacic, J., *The inverse problem in the Galois theory of differential fields*, Ann. of Math., **89** (1969), 1151-1164.

[35] Larson, R.G. and Taft, E.J., *The algebraic structure of linearly recursive sequences under Hadamard product*, Isreal J. Math., **72** (1990), 118-132.

[36] Levelt, A.H.M., *Differential Galois theory and tensor products*, Indag. Math., New Series 1(4) (1990), 439-450.

[37] Loday-Richaud, M., *Stokes phenomenon, multisummability and differential Galois groups*, Ann. Inst. Fourier, **44** (1994), 849-906.

[38] Magid, A., *Finite generation of class groups of rings of invariants*, Proc. Amer. Math.Soc., **60** (1976), 45-48.

[39] Magid, A., **Lectures on Differential Galois Theory**, University Lecture Series, Vol. 7, American Mathematical Society, (1994).

[40] Martinet, J and Ramis, J.-P., *Elementary acceleration and multisummability*, Ann. de l'I.H.P. 54 (1991), 331-401.

[41] Mitschi, C., Singer, M.F., *Connected linear algebraic groups as Galois groups*, J. of Algebra, 184(1996), 333-361.

[42] Nörlund N.E., **Leçons sur les séries d'Interpolation**, Gauthiers Villars et Cie, Paris 1926.

[43] Petkovšek, M., *Finding closed form solutions of difference equations by symbolic methods*, thesis, Department of Computer Science, Carnegie Mellon University, 1990.

[44] Petkovšek, M., *Hypergeometric solutions of linear recurrences with polynomial coefficients*, J. Symb. Comp., **14** (1992), 243-264.

[45] Petkovšek, M., *A generalization of Gosper's algorithm*, Discrete Mathematics, **134** (1994), 125-131.

[46] Petkovšek, M., Wilf, H., Zeilberger, D., **A=B**, A.K. Peters, Wellsely, Massachusets, 1996.

[47] Praagman, C., *The formal classification of linear difference operators*, Proc. Kon. Ned. Ac. Wet. Ser A, **86** (1983), 249-261.

[48] Praagman, C., *Meromorphic linear difference equations*, Thesis, University of Groningen, 1985.

[49] van der Put, M., *Galoistheorie van differentiaalvergelijkingen* unpublished lecture notes (in Dutch), Mathematisch Instituut, University of Groningen, The Netherlands, 1984.

[50] van der Put, M., *Differential equations in characteristic p*, Compositio Mathematica **97** (1995), 227-251.

[51] van der Put, M., *Singular Complex Differential Equations: An Introduction*, Nieuw Archief voor Wiskunde, vierde serie, deel **13** No. 3 (1995), 451-470.

[52] Ramis, J.P., *About the growth of entire functions solutions of linear algebraic q-difference equations* Annales de Fac. des Sciences de Toulouse, Série 6, Vol. 1, no 1 (1992), 53-94.

[53] Ramis, J.P., *About the inverse problem in differential Galois theory: The differential Abhyankar conjecture*, in "The Stoke Phenomenon and Hilbert's 16th Problem", World Scientific Publishers, 1996 (editors B.L.J.Braaksma, G.K. Immink, M. van der Put)

[54] Renault, G., **Algèbre non commutative**, Gautier-Villars, Paris, 1975.

[55] Reutenauer, C., *Sur les éléments inversibles de l'algèbre de Hadamard des séries rationelles*, Bull. Soc. Math. France, **110** (1982), 225-232.

[56] Rosenlicht, M., *Toroidal algebraic groups*, Proc. Amer. Math. Soc., **12**, (1961), 984-988.

[57] Serre, J.P., **Cohomologie Galoisienne**, Lecture Notes in Mathematics, Springer-Verlag, New York, 1964.

[58] Serre, J.P., **Corps Locaux**, Hermann, Paris, 1962.

[59] Singer, M.F., *Algebraic relations among solutions of linear differential equations*, Trans. Am. Math. Soc., **295**(2) (1986) 753-763.

[60] Tretkoff, C., and Tretkoff, M., *Solution of the inverse problem of differential Galois theory in the classical case*, Amer. J. Math., **101** (1979), 1327-1332.

[61] Varadarajan, V.S., *Meromorphic differential equations*, Expo.Math. **9**, (1991), 97-188.

[62] Whittaker, E.T., Watson, G.N., **A Course of Modern Analysis**, Fourth Edition, Cambridge University Press, 1978.

[63] Zariski, O. and Samuel, P, **Commutative Algebra**, Vol. 1, D. Nostrand, Princeton, NJ, 1958.

Index

Notations

Lecture Notes in Mathematics

For information about Vols. 1–1479
please contact your bookseller or Springer-Verlag

Vol. 1480: F. Dumortier, R. Roussarie, J. Sotomayor, H. Żołądek, Bifurcations of Planar Vector Fields: Nilpotent Singularities and Abelian Integrals. VIII, 226 pages. 1991.

Vol. 1481: D. Ferus, U. Pinkall, U. Simon, B. Wegner (Eds.), Global Differential Geometry and Global Analysis. Proceedings, 1991. VIII, 283 pages. 1991.

Vol. 1482: J. Chabrowski, The Dirichlet Problem with L^2-Boundary Data for Elliptic Linear Equations. VI, 173 pages. 1991.

Vol. 1483: E. Reithmeier, Periodic Solutions of Nonlinear Dynamical Systems. VI, 171 pages. 1991.

Vol. 1484: H. Delfs, Homology of Locally Semialgebraic Spaces. IX, 136 pages. 1991.

Vol. 1485: J. Azéma, P. A. Meyer, M. Yor (Eds.), Séminaire de Probabilités XXV. VIII, 440 pages. 1991.

Vol. 1486: L. Arnold, H. Crauel, J.-P. Eckmann (Eds.), Lyapunov Exponents. Proceedings, 1990. VIII, 365 pages. 1991.

Vol. 1487: E. Freitag, Singular Modular Forms and Theta Relations. VI, 172 pages. 1991.

Vol. 1488: A. Carboni, M. C. Pedicchio, G. Rosolini (Eds.), Category Theory. Proceedings, 1990. VII, 494 pages. 1991.

Vol. 1489: A. Mielke, Hamiltonian and Lagrangian Flows on Center Manifolds. X, 140 pages. 1991.

Vol. 1490: K. Metsch, Linear Spaces with Few Lines. XIII, 196 pages. 1991.

Vol. 1491: E. Lluis-Puebla, J.-L. Loday, H. Gillet, C. Soulé, V. Snaith, Higher Algebraic K-Theory: an overview. IX, 164 pages. 1992.

Vol. 1492: K. R. Wicks, Fractals and Hyperspaces. VIII, 168 pages. 1991.

Vol. 1493: E. Benoît (Ed.), Dynamic Bifurcations. Proceedings, Luminy 1990. VII, 219 pages. 1991.

Vol. 1494: M.-T. Cheng, X.-W. Zhou, D.-G. Deng (Eds.), Harmonic Analysis. Proceedings, 1988. IX, 226 pages. 1991.

Vol. 1495: J. M. Bony, G. Grubb, L. Hörmander, H. Komatsu, J. Sjöstrand, Microlocal Analysis and Applications. Montecatini Terme, 1989. Editors: L. Cattabriga, L. Rodino. VII, 349 pages. 1991.

Vol. 1496: C. Foias, B. Francis, J. W. Helton, H. Kwakernaak, J. B. Pearson, H_∞-Control Theory. Como, 1990. Editors: E. Mosca, L. Pandolfi. VII, 336 pages. 1991.

Vol. 1497: G. T. Herman, A. K. Louis, F. Natterer (Eds.), Mathematical Methods in Tomography. Proceedings 1990. X, 268 pages. 1991.

Vol. 1498: R. Lang, Spectral Theory of Random Schrödinger Operators. X, 125 pages. 1991.

Vol. 1499: K. Taira, Boundary Value Problems and Markov Processes. IX, 132 pages. 1991.

Vol. 1500: J.-P. Serre, Lie Algebras and Lie Groups. VII, 168 pages. 1992.

Vol. 1501: A. De Masi, E. Presutti, Mathematical Methods for Hydrodynamic Limits. IX, 196 pages. 1991.

Vol. 1502: C. Simpson, Asymptotic Behavior of Monodromy. V, 139 pages. 1991.

Vol. 1503: S. Shokranian. The Selberg-Arthur Trace Formula (Lectures by J. Arthur). VII, 97 pages. 1991.

Vol. 1504: J. Cheeger, M. Gromov, C. Okonek, P. Pansu, Geometric Topology: Recent Developments. Editors: P. de Bartolomeis, F. Tricerri. VII, 197 pages. 1991.

Vol. 1505: K. Kajitani, T. Nishitani, The Hyperbolic Cauchy Problem. VII, 168 pages. 1991.

Vol. 1506: A. Buium, Differential Algebraic Groups of Finite Dimension. XV, 145 pages. 1992.

Vol. 1507: K. Hulek, T. Peternell, M. Schneider, F.-O. Schreyer (Eds.), Complex Algebraic Varieties. Proceedings, 1990. VII, 179 pages. 1992.

Vol. 1508: M. Vuorinen (Ed.), Quasiconformal Space Mappings. A Collection of Surveys 1960-1990. IX, 148 pages. 1992.

Vol. 1509: J. Aguadé, M. Castellet, F. R. Cohen (Eds.), Algebraic Topology - Homotopy and Group Cohomology. Proceedings, 1990. X, 330 pages. 1992.

Vol. 1510: P. P. Kulish (Ed.), Quantum Groups. Proceedings, 1990. XII, 398 pages. 1992.

Vol. 1511: B. S. Yadav, D. Singh (Eds.), Functional Analysis and Operator Theory. Proceedings, 1990. VIII, 223 pages. 1992.

Vol. 1512: L. M. Adleman, M.-D. A. Huang, Primality Testing and Abelian Varieties Over Finite Fields. VII, 142 pages. 1992.

Vol. 1513: L. S. Block, W. A. Coppel, Dynamics in One Dimension. VIII, 249 pages. 1992.

Vol. 1514: U. Krengel, K. Richter, V. Warstat (Eds.), Ergodic Theory and Related Topics III, Proceedings, 1990. VIII, 236 pages. 1992.

Vol. 1515: E. Ballico, F. Catanese, C. Ciliberto (Eds.), Classification of Irregular Varieties. Proceedings, 1990. VII, 149 pages. 1992.

Vol. 1516: R. A. Lorentz, Multivariate Birkhoff Interpolation. IX, 192 pages. 1992.

Vol. 1517: K. Keimel, W. Roth, Ordered Cones and Approximation. VI, 134 pages. 1992.

Vol. 1518: H. Stichtenoth, M. A. Tsfasman (Eds.), Coding Theory and Algebraic Geometry. Proceedings, 1991. VIII, 223 pages. 1992.

Vol. 1519: M. W. Short, The Primitive Soluble Permutation Groups of Degree less than 256. IX, 145 pages. 1992.